DATE DUE

GAYLORD			PRINTED IN U.S.A.

THE LAST
RECREATIONS

THE LAST RECREATIONS

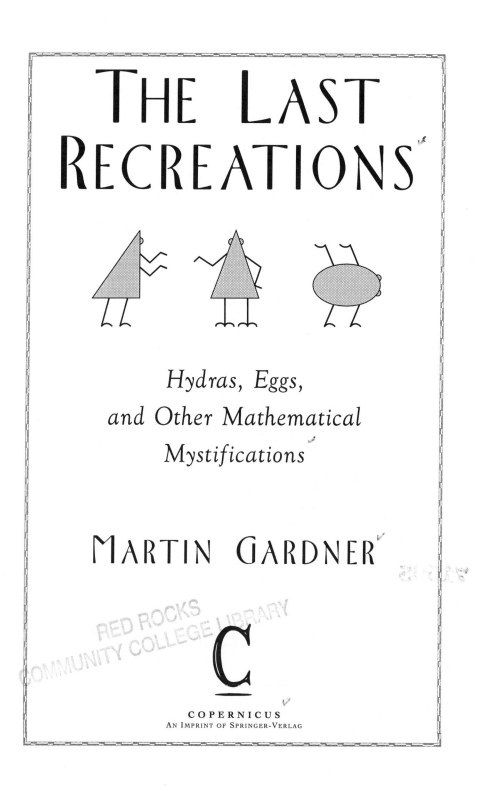

Hydras, Eggs,
and Other Mathematical
Mystifications

MARTIN GARDNER

C

COPERNICUS
AN IMPRINT OF SPRINGER-VERLAG

Published in the United States by Copernicus, an imprint of
Springer-Verlag New York, Inc.

Copernicus
Springer-Verlag New York, Inc.
175 Fifth Avenue
New York, NY 10010
USA

Library of Congress Cataloging-in-Publication Data
Gardner, Martin, 1914–
   The last recreations: hydras, eggs, and other mathematical
   mystifications/Martin Gardner.
      p.   cm.
   Includes index.
   ISBN 0-387-94929-1 (hardcover: alk. paper)
   1. Mathematical recreations.   I. Title.
QA95.G255 1997
793.7'4–DC21   96-51641

Manufactured in the United States of America.
Printed on acid-free paper.
Designed by Irmgard Lochner.

9 8 7 6 5 4 3 2 1

ISBN 0-387-94929-1   SPIN 10561414

*To Persi Diaconis for his remarkable
contributions to mathematics and conjuring,
for his unswerving opposition to psychic
nonsense, and for a friendship going back to our
Manhattan years.*

Preface

One of my greatest pleasures and privileges was writing a column, over a period of some 30 years, for *Scientific American*. It began with an article on hexaflexagons in December 1956 and concluded with a column on minimal Steiner trees in May 1986.

Writing this column was a marvelous learning experience. I took no courses in math when I was an undergraduate at the University of Chicago—my major was philosophy—but I have always loved mathematics, and now and then regret I did not pursue it as a career. It takes only a glance through earlier book collections of the columns to see how they gradually became more sophisticated as I learned more about mathematics. Not the least of my delights was getting to know many truly eminent mathematicians who generously contributed material and who have since become lifelong friends.

This is the fifteenth and final collection. As in previous books in the series, I have done my best to correct blunders, to expand and update each column with an addendum, to add new illustrations, and to provide fuller lists of selected references.

Martin Gardner

Contents

1

The Wonders of a Planiverse

"Planiversal scientists are not a very common breed."

—Alexander Keewatin Dewdney

A s far as anyone knows the only existing universe is the one we live in, with its three dimensions of space and one of time. It is not hard to imagine, as many science-fiction writers have, that intelligent organisms could live in a four-dimensional space, but two dimensions offer such limited degrees of freedom that it has long been assumed intelligent two-space life forms could not exist. Two notable attempts have nonetheless been made to describe such organisms.

In 1884 Edwin Abbott Abbott, a London clergyman, published his satirical novel *Flatland*. Unfortunately the book leaves the reader almost entirely in the dark about Flatland's physical laws and the technology developed by its inhabitants, but the situation was greatly improved in 1907 when Charles Howard Hinton published *An Episode of Flatland*. Although written in a flat style and with cardboard characters, Hinton's story provided the first glimpses of the possible science and technology of the two-dimensional world. His eccentric book is, alas, long out of print, but you can read about it in the chapter "Flatlands" in my book *The Unexpected Hanging and Other Mathematical Diversions* (Simon & Schuster, 1969).

In "Flatlands" I wrote: "It is amusing to speculate on two-dimensional physics and the kinds of simple mechanical devices that would be feasible in a flat world." This remark caught the attention of Alexander Keewatin Dewdney, a computer scientist at the University of Western Ontario. Some of his early speculations on the subject were set down in 1978 in a university report and in 1979 in "Exploring the Planiverse," an article in *Journal of Recreational Mathematics* (Vol. 12, No. 1, pages 16–20; September). Later in 1979 Dewdney also privately published "Two-dimensional Science and Technology," a 97-page tour de force. It is hard to believe, but Dewdney actually lays the groundwork for what he calls a planiverse: a possible two-dimensional world. Complete with its own laws of chemistry, physics, astronomy, and biology, the planiverse is closely analogous to our own universe (which he calls the steriverse) and is apparently free of contradictions. I should add that this remarkable achievement is an amusing hobby for a mathematician whose serious contributions have appeared in some 30 papers in technical journals.

Dewdney's planiverse resembles Hinton's in having an earth that he calls (as Hinton did) Astria. Astria is a disklike planet that rotates in planar space. The Astrians, walking upright on the rim of the planet, can distinguish east and west and up and down. Naturally there is no north or south. The "axis" of Astria is a point at the center of the circular planet. You can think of such a flat planet as being truly two-dimensional or you can give it a very slight thickness and imagine it as moving between two frictionless planes.

As in our world, gravity in a planiverse is a force between objects that varies directly with the product of their masses, but it varies inversely with the *linear* distance between them, not with the square of that distance. On the assumption that forces such as light and gravity in a planiverse move in straight lines, it is easy to see that the intensity of such forces must vary inversely with linear distance. The familiar text-book figure demonstrating that in our world the intensity of light varies inversely with the square of distance is shown at the top of Figure 1. The obvious planar analogue is shown at the bottom of the illustration.

Figure 1

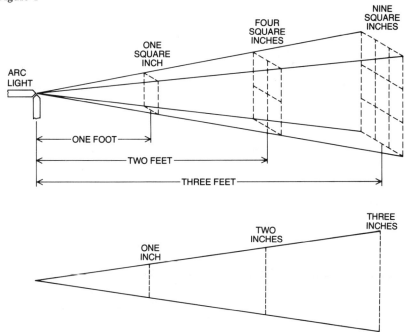

To keep his whimsical project from "degenerating into idle specula-
tion" Dewdney adopts two basic principles. The "principle of similarity" states that the planiverse must be as much like the steriverse as
possible: a motion not influenced by outside forces follows a straight
line, the flat analogue of a sphere is a circle, and so on. The "principle
of modification" states that in those cases where one is forced to choose
between conflicting hypotheses, each one equally similar to a steriversal
theory, the more fundamental one must be chosen and the other modified. To determine which hypothesis is more fundamental Dewdney
relies on the hierarchy in which physics is more fundamental than
chemistry, chemistry more fundamental than biology, and so on.

To illustrate the interplay between levels of theory Dewdney considers the evolution of the planiversal hoist in Figure 2. The engineer who
designed it first gave it arms thinner than those in the illustration, but
when a metallurgist pointed out that planar materials fracture more
easily than their three-space counterparts, the engineer made the arms

Figure 2

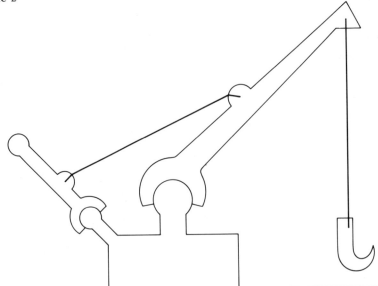

thicker. Later a theoretical chemist, invoking the principles of similarity and modification at a deeper level, calculated that the planiversal molecular forces are much stronger than had been suspected, and so the engineer went back to thinner arms.

The principle of similarity leads Dewdney to posit that the planiverse is a three-dimensional continuum of space–time containing matter composed of molecules, atoms, and fundamental particles. Energy is propagated by waves, and it is quantized. Light exists in all its wavelengths and is refracted by planar lenses, making possible planiversal eyes, planiversal telescopes, and planiversal microscopes. The planiverse shares with the steriverse such basic precepts as causality; the first and second laws of thermodynamics; and laws concerning inertia, work, friction, magnetism, and elasticity.

Dewdney assumes that his planiverse began with a big bang and is currently expanding. An elementary calculation based on the inverse-linear gravity law shows that regardless of the amount of mass in the planiverse the expansion must eventually halt, so that a contracting phase will begin. The Astrian night sky will of course be a semicircle along which are scattered twinkling points of light. If the stars have proper motions, they will continually be occulting one another. If Astria has a sister planet, it will over a period of time occult every star in the sky.

We can assume that Astria revolves around a sun and rotates, thereby creating day and night. In a planiverse, Dewdney discovered, the only stable orbit that continually retraces the same path is a perfect circle. Other stable orbits roughly elliptical in shape are possible, but the axis of the ellipse rotates in such a way that the orbit never exactly closes. Whether planiversal gravity would allow a moon to have a stable orbit around Astria remains to be determined. The difficulty is due to the sun's gravity, and resolving the question calls for work on the planar analogue of what our astronomers know as the three-body problem.

Dewdney analyzes in detail the nature of Astrian weather, using analogies to our seasons, winds, clouds, and rain. An Astrian river would be indistinguishable from a lake except that it might have faster currents. One peculiar feature of Astrian geology is that water cannot flow around a rock as it does on the earth. As a result rainwater steadily accumulates behind any rock on a slope, tending to push the rock downhill: the gentler the slope is, the more water accumulates and the stronger the push is. Dewdney concludes that given periodic rainfall the Astrian surface would be unusually flat and uniform. Another consequence of the inability of water to move sideways on Astria is that it would become trapped in pockets within the soil, tending to create large areas of treacherous quicksand in the hollows of the planet. One hopes, Dewdney writes, that rainfall is infrequent on Astria. Wind too would have much severer effects on Astria than on the earth because like rain it cannot "go around" objects.

Dewdney devotes many pages to constructing a plausible chemistry for his planiverse, modeling it as much as possible on three-dimensional matter and the laws of quantum mechanics. Figure 3 shows Dewdney's periodic table for the first 16 planiversal elements. Because the first two are so much like their counterparts in our world, they are called hydrogen and helium. The next 10 have composite names to suggest the steriversal elements they most resemble; for example, lithrogen combines the properties of lithium and nitrogen. The next four are named after Hinton, Abbott, and the young lovers in Hinton's novel, Harold Wall and Laura Cartwright.

In the flat world atoms combine naturally to form molecules, but of course only bonding that can be diagrammed by a planar graph is allowed. (This result follows by analogy from the fact that intersecting bonds do not exist in steriversal chemistry.) As in our world, two asymmetric molecules can be mirror images of each other, so that neither

ATOMIC NUMBER	NAME	SYMBOL	SHELL STRUCTURE								VALENCE
			1s	2s	2p	3s	3p	3d	4s	4p ...	
1	HYDROGEN	H	1								1
2	HELIUM	He	2								2
3	LITROGEN	Lt	2	1							1
4	BEROXYGEN	Bx	2	2							2
5	FLUORON	Fl	2	2	1						3
6	NEOCARBON	Nc	2	2	2						4
7	SODALINUM	Sa	2	2	2	1					1
8	MAGNILICON	Mc	2	2	2	2					2
9	ALUPHORUS	Ap	2	2	2	2	1				3
10	SULFICON	Sp	2	2	2	2	2				4
11	CHLOPHORUS	Cp	2	2	2	2	2	1			5
12	ARGOFUR	Af	2	2	2	2	2	2			6
13	HINTONIUM	Hn	2	2	2	2	2	2	1		1
14	ABBOGEN	Ab	2	2	2	2	2	2	2		2
15	HAROLDIUM	Wa	2	2	2	2	2	2	2	1	3
16	LAURANIUM	La	2	2	2	2	2	2	2	2	4

Figure 3

one can be "turned over" to become identical with the other. There are striking parallels between planiversal chemistry and the behavior of steriversal monolayers on crystal surfaces [see "Two-dimensional Matter," by J. G. Dash; *Scientific American*, May 1973]. In our world molecules can form 230 distinct crystallographic groups, but in the planiverse they can form only 17. I am obliged to pass over Dewdney's speculations about the diffusion of molecules, electrical and magnetic laws, analogues of Maxwell's equations, and other subjects too technical to summarize here.

Dewdney assumes that animals on Astria are composed of cells that cluster to form bones, muscles, and connective tissues similar to those found in steriversal biology. He has little difficulty showing how these bones and muscles can be structured to move appendages in such a way that the animals can crawl, walk, fly, and swim. Indeed, some of these movements are easier in a planiverse than in our world. For example, a steriversal animal with two legs has considerable difficulty balancing while walking, whereas in the planiverse if an animal has both legs on the ground, there is no way it can fall over. Moreover, a flying planiversal animal cannot have wings and does not need them to fly; if the body of the animal is aerodynamically shaped, it can act as a wing (since air can go around it only in the plane). The flying animal could be propelled by a flapping tail.

Calculations also show that Astrian animals probably have much lower metabolic rates than terrestrial animals because relatively little heat is lost through the perimeter of their body. Furthermore, animal bones can be thinner on Astria than they are on the earth, because they have less weight to support. Of course, no Astrian animal can have an open tube extending from its mouth to its anus, because if it did, it would be cut in two.

In the appendix to his book *The Structure and Evolution of the Universe* (Harper, 1959) G. J. Whitrow argues that intelligence could not evolve in two-space because of the severe restrictions two dimensions impose on nerve connections. "In three or more dimensions," he writes, "any number of [nerve] cells can be connected with [one another] in pairs without intersection of the joins, but in two dimensions the maximum number of cells for which this is possible is only four." Dewdney easily demolishes this argument, pointing out that if nerve cells are allowed to fire nerve impulses through "crossover points," they can form flat networks as complex as any in the steriverse. Planiversal minds would operate more slowly than steriversal ones, however, because in the two-dimensional networks the pulses would encounter more interruptions. (There are comparable results in the theory of two-dimensional automatons.)

Dewdney sketches in detail the anatomy of an Astrian female fish with a sac of unfertilized eggs between its two tail muscles. The fish has an external skeleton, and nourishment is provided by the internal circulation of food vesicles. If a cell is isolated, food enters it through a membrane that can have only one opening at a time. If the cell is in contact with other cells, as in a tissue, it can have more than one opening at a time because the surrounding cells are able to keep it intact. We can of course see every internal organ of the fish or of any other planiversal life form, just as a four-dimensional animal could see all our internal organs.

Dewdney follows Hinton in depicting his Astrian people schematically, as triangles with two arms and two legs. Hinton's Astrians, however, always face in the same direction: males to the east and females to the west. In both sexes the arms are on the front side, and there is a single eye at the top of the triangle, as shown in Figure 4. Dewdney's Astrians are bilaterally symmetrical, with an arm, a leg, and an eye on each side, as shown in the illustration's center. Hence these Astrians, like terrestrial birds or horses, can see in opposite directions. Naturally the only way for one Astrian to pass another is to crawl or leap over him. My conception of an Astrian bug-eyed monster is shown at the right in the illustration. This creature's appendages serve as either arms or legs, depending on which way it is facing, and its two eyes provide binocular vision. With only one eye an Astrian would have a largely one-dimensional visual world, giving him a rather narrow perception of reality. On the other hand, parts of objects in the planiverse might be distinguished by their color, and an illusion of depth might be created by the focusing of the lens of the eye.

On Astria building a house or mowing a lawn requires less work

Figure 4

than it does on the earth because the amount of material involved is considerably smaller. As Dewdney points out, however, there are still formidable problems to be dealt with in a two-dimensional world: "Assuming that the surface of the planet is absolutely essential to support life-giving plants and animals, it is clear that very little of the Astrian surface can be disturbed without inviting the biological destruction of the planet. For example, here on earth we may build a modest highway through the middle of several acres of rich farmland and destroy no more than a small percentage of it. A corresponding highway on Astria with destroy *all* the 'acreage' it passes over. . . . Similarly, extensive cities would quickly use up the Astrian countryside. It would seem that the only alternative for the Astrian technological society is to go underground." A typical subterranean house with a living room, two bedrooms, and a storage room is shown in Figure 5. Collapsible chairs and tables are stored in recesses in the floors to make the rooms easier to walk through.

The many simple three-dimensional mechanical elements that have obvious analogues on Astria include rods, levers, inclined planes, springs, hinges, ropes, and cables (see Figure 6, top). Wheels can be rolled along the ground, but there is no way to turn them on a fixed axle. Screws are impossible. Ropes cannot be knotted; but by the same token, they never tangle. Tubes and pipes must have partitions, to keep their sides in place, and the partitions have to be opened (but never all of them at once) to allow anything to pass through. It is remarkable that in spite of these severe constraints many flat mechanical devices can be built that will work. A faucet designed by Dewdney is shown in Figure 6, bottom. To operate it the handle is lifted. This action pulls the valve away from the wall of the spout, allowing the water to flow out. When the handle is released, the spring pushes the valve back.

The device shown in Figure 7 serves to open and close a door (or a wall). Pulling down the lever at the right forces the wedge at the

Figure 5

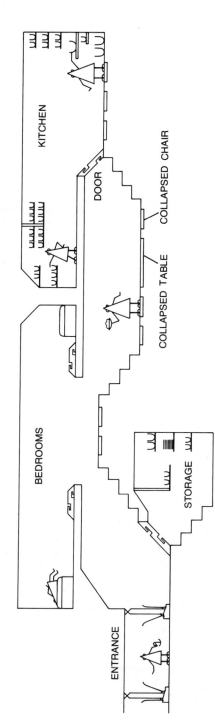

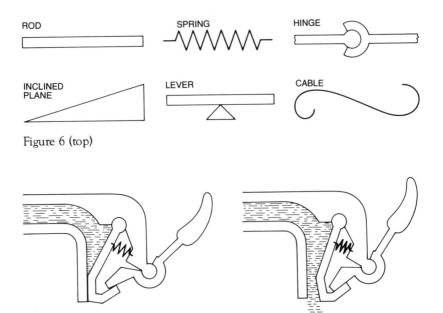

Figure 6 (top)

Figure 6 (bottom)

bottom to the left, thereby allowing the door to swing upward (carrying the wedge and the levers with it) on a hinge at the top. The door is opened from the left by pushing up on the other lever. The door can be lowered from either side and the wedge moved back to stabilize the wall by moving a lever in the appropriate direction. This device and the faucet are both mechanisms with permanent planiversal hinges: circular knobs that rotate inside hollows but cannot be removed from them.

Figure 8 depicts a planiversal steam engine whose operation parallels that of a steriversal engine. Steam under pressure is admitted into the cylinder of the engine through a sliding valve that forms one of its walls (*top*). The steam pressure causes a piston to move to the right until steam can escape into a reservoir chamber above it. The subsequent loss of pressure allows the compound leaf spring at the right of the cylinder to drive the piston back to the left (*bottom*). The sliding valve is

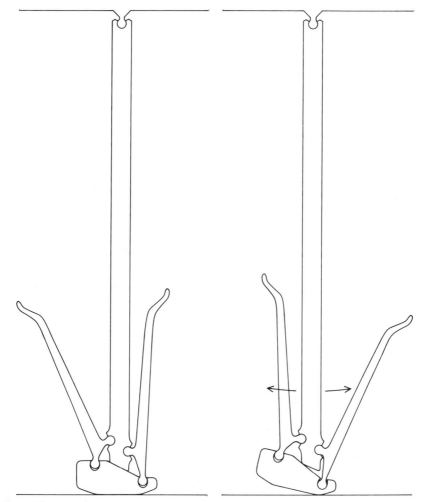

Figure 7

closed as the steam escapes into the reservoir, but as the piston moves back it reopens, pulled to the right by a spring-loaded arm.

Figure 9 depicts Dewdney's ingenious mechanism for unlocking a door with a key. This planiversal lock consists of three slotted tumblers (a) that line up when a key is inserted (b) so that their lower halves move as a unit when the key is pushed (c). The pushing of the key is transmitted through a lever arm to the master latch, which pushes

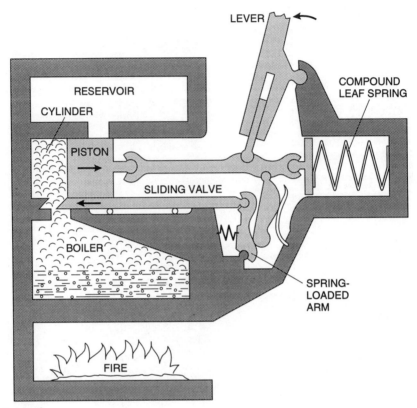

Figure 8

down on a slave latch until the door is free to swing to the right (d). The bar on the lever arm and the lip on the slave latch make the lock difficult to pick. Simple and compound leaf springs serve to return all the parts of the lock except the lever arm to their original positions when the door is opened and the key is removed. When the door closes, it strikes the bar on the lever arm, thereby returning that piece to its original position as well. This flat lock could actually be employed in the steriverse; one simply inserts a key without twisting it.

"It is amusing to think," writes Dewdney, "that the rather exotic design pressures created by the planiversal environment could cause us to think about mechanisms in such a different way that entirely novel solutions to old problems arise. The resulting designs, if steriversally practical, are invariably space-saving."

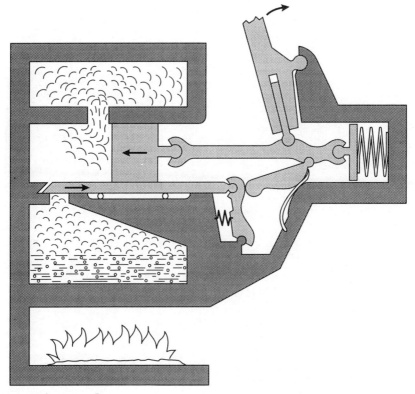

Figure 8 (continued)

Thousands of challenging planiversal problems remain unsolved. Is there a way, Dewdney wonders, to design a two-dimensional windup motor with flat springs or rubber bands that would store energy? What is the most efficient design for a planiversal clock, telephone, book, typewriter, car, elevator or computer? Will some machines need a substitute for the wheel and axle? Will some need electric power?

There is a curious pleasure in trying to invent machines for what Dewdney calls "a universe both similar to and yet strangely different from ours." As he puts it, "from a small number of assumptions many phenomena seem to unfurl, giving one the sense of a kind of separate existence of this two-dimensional world. One finds oneself speaking, willy-nilly, of *the* planiverse as opposed to *a* planiverse. . . . [For] those

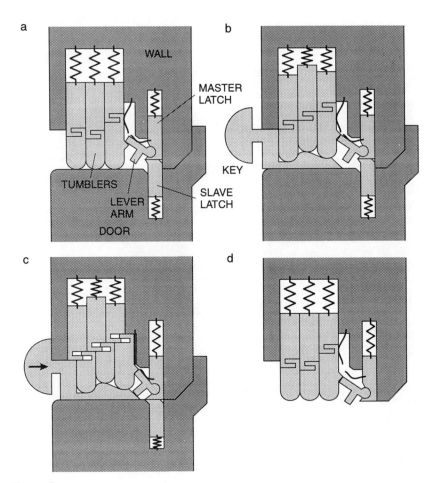

Figure 9

who engage in it positively, there is a kind of strange enjoyment, like
[that of] an explorer who enters a land where his own perceptions play
a major role in the landscape that greets his eyes."

Some philosophical aspects of this exploration are not trivial. In
constructing a planiverse one sees immediately that it cannot be built
without a host of axioms that Leibniz called the "compossible" elements of any possible world, elements that allow a logically consistent
structure. Yet as Dewdney points out, science in our universe is based
mainly on observations and experiments, and it is not easy to find any

underlying axioms. In constructing a planiverse we have nothing to observe. We can only perform *gedanken* experiments (thought experiments) about what might be observed. "The experimentalist's loss," observes Dewdney, "is the theoretician's gain."

A marvelous exhibit could be put on of working models of planiversal machines, cut out of cardboard or sheet metal, and displayed on a surface that slopes to simulate planiversal gravity. One can also imagine beautiful cardboard exhibits of planiversal landscapes, cities, and houses. Dewdney has opened up a new game that demands knowledge of both science and mathematics: the exploration of a vast fantasy world about which at present almost nothing is known.

It occurs to me that Astrians would be able to play two-dimensional board games but that such games would be as awkward for them as three-dimensional board games are for us. I imagine them, then, playing a variety of linear games on the analogue of our 8-by-8 chessboard. Several games of this type are shown in Figure 10. Part *a* shows the start of a checkers game. Pieces move forward only, one cell at a time, and jumps are compulsory. The linear game is equivalent to a game of regular checkers with play confined to the main diagonal of a standard board. It is easy to see how the second player wins in rational play and how in misère, or "giveaway," checkers the first player wins just as easily. Linear checkers games become progressively harder to analyze as longer boards are introduced. For example, which player wins standard linear checkers on the 11-cell board when each player starts with checkers on the first four cells at his end of the board?

Part *b* in the illustration shows an amusing Astrian analogue of chess. On a linear board a bishop is meaningless and a queen is the same as a rook, so the pieces are limited to kings, knights, and rooks. The only rule modification needed is that a knight moves two cells in either direction and can jump an intervening piece of either color. If

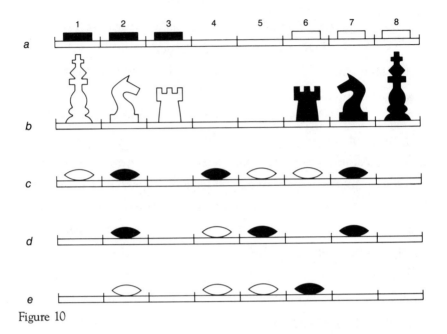

Figure 10

the game is played rationally, will either White or Black win or will the game end in a draw? The question is surprisingly tricky to answer.

Linear go, played on the same board, is by no means trivial. The version I shall describe was invented 10 years ago by James Marston Henle, a mathematician who is now at Smith College. Called pinch by Henle, it is published here for the first time.

In the game of pinch players take turns placing black and white stones on the cells of the linear board, and whenever the stones of one player surround the stones of the other, the surrounded stones are removed. For example, both sets of white stones shown in part c of Figure 10 are surrounded. Pinch is played according to the following two rules.

Rule 1: No stone can be placed on a cell where it is surrounded unless that move serves to surround a set of enemy stones. Hence in the situation shown in part d of the illustration White cannot play on cells 1, 3, or 8, but he can play on cell 6 because this move serves to surround cell 5.

Rule 2: A stone cannot be placed on a cell from which a stone was removed on the last play if the purpose of the move is to surround something. A player must wait at least one turn before making such a move. For example, in part *e* of the illustration assume that Black plays on cell 3 and removes the white stones on cells 4 and 5. White cannot play on cell 4 (to surround cell 3) for his next move, but he may do so for any later move. He can play on cell 5, however, because even though a stone was just removed from that cell, the move does not serve to surround anything. This rule is designed to decrease the number of stalemates, as is the similar rule in go.

Two-cell pinch is a trivial win for the second player. The three- and four-cell games are easy wins for the first player if he takes the center in the three-cell game and one of the two central cells in the four-cell one. The five-cell game is won by the second player and the six- and seven-cell games are won by the first player. The eight-cell game jumps to such a high level of complexity that it becomes very exciting to play. Fortunes often change rapidly, and in most situations the winning player has only one winning move.

Answers

In 11-cell linear checkers (beginning with Black on cells 1, 2, 3, and 4 and White on cells 8, 9, 10, and 11) the first two moves are forced: Black to 5 and White to 7. To avoid losing, Black then goes to 4, and White must respond by moving to 8. Black is then forced to 3 and White to 9. At this point Black loses with a move to 2 but wins with a move to 6. In the latter case White jumps to 5, and then Black jumps to 6 for an easy end-game victory.

On the eight-cell linear chessboard White can win in at most six moves. Of White's four opening moves, R×R is an instant stalemate and the shortest possible game. R-5 is a quick loss for White if Black plays R×R. Here White must respond with N-4, and then Black mates on his second move with R×N. This game is one of the two "fool's mates," or shortest possible wins. The R-4 opening allows Black to mate on his second or third move if he responds with N-5.

White's only winning opening is N-4. Here Black has three possible replies:

1. R×N.

In this case White wins in two moves with R×R.

2. R-5.

White wins with K-2. If Black plays R-6, White mates with R×N. If Black takes the knight, White takes the rook, Black moves N-5, and White mates by taking Black's knight.

3. N-5.

This move delays Black's defeat the longest. In order to win White must check with N×R, forcing Black's king to 7. White moves his rook to 4. If Black plays K×N, White's king goes to 2, Black's K-7 is forced, and White's R×N wins. If Black plays N-3 (check), White moves the king to 2. Black can move only the knight. If he plays N-1, White mates with N-8. If Black plays N-5, White's N-8 forces Black's K×N, and then White mates with R×N.

The first player also has the win in eight-cell pinch (linear go) by opening on the second cell from an end, a move that also wins the six- and seven-cell games. Assume that the first player plays on cell 2. His unique winning responses to his opponent's plays on 3, 4, 5, 6, 7, and 8 are respectively 5, 7, 7, 7, 5, and 6. I leave the rest of the game to the reader. It is not known whether there are other winning opening moves. James Henle, the inventor of pinch, assures me that the second player wins the nine-cell game. He has not tried to analyze boards with more than nine cells.

ADDENDUM

My column on the planiverse generated enormous interest. Dewdney received some thousand letters offering suggestions about flatland science and technology. In 1979 he privately printed *Two-Dimensional Science and Technology*, a monograph discussing these new results. Two years later he edited another monograph, *A Symposium of Two-Dimensional Science and Technology*. It contained papers by noted scientists, mathematicians, and laymen, grouped under the categories of physics, chemistry, astronomy, biology, and technology. *Newsweek* covered these monographs in a two-page article, "Life in Two Dimensions" (January 18, 1980), and a similar article, "Scientific Dreamers' Worldwide Cult," ran in Canada's *Maclean's* magazine (January 11, 1982). *Omni* (March 1983), in an article on "Flatland Redux," included a photograph of Dewdney shaking hands with an Astrian.

In 1984 Dewdney pulled it all together in a marvelous work, half nonfiction and half fantasy, titled *The Planiverse* and published by Poseidon Press, an imprint of Simon & Schuster. That same year he took over the mathematics column in *Scientific American*, shifting its emphasis to computer recreations. Several collections of his columns have been published by W. H. Freeman: *The Armchair Universe* (1987), *The Turing Omnibus* (1989), and *The Magic Machine* (1990).

An active branch of physics is now devoted to planar phenomena. It involves research on the properties of surfaces covered by a film one molecule thick, and a variety of two-dimensional electrostatic and electronic effects. Exploring possible flatlands also relates to a philosophical fad called "possible worlds." Extreme proponents of this movement actually argue that if a universe is logically possible—that is, free of logical contradictions—it is just as "real" as the universe in which we flourish.

In *Childhood's End* Arthur Clarke describes a giant planet where intense gravity has forced life to evolve almost flat forms with a vertical thickness of one centimeter.

The following letter from J. Richard Gott III, an astrophysicist at Princeton University, was published in *Scientific American* (October 1980):

> I was interested in Martin Gardner's article on the physics of Flatland, because for some years I have given the students in my general relativity class the problem of deriving the theory of general relativity for Flatland. The results are surprising. One does *not* obtain the Flatland analogue of Newtonian theory (masses with gravitational fields falling off like $1/r$) as the weak-field limit. General relativity in Flatland predicts no gravitational waves and no action at a distance. A planet in Flatland would produce no gravitational effects beyond its own radius. In our four-dimensional space–time the energy momentum tensor has 10 independent components, whereas the Riemann curvature tensor has 20 independent components. Thus it is possible to find solutions to the vacuum field equations $G_{\mu\nu} = 0$ (where all components of the energy momentum tensor are zero) that have a nonzero curvature. Black-hole solutions and the gravitational-field solution external to a planet are examples. This allows gravitational waves and action at a distance. Flatland has a three-dimensional space–time where the energy momentum tensor has six independent components and the Riemann curvature tensor also has only six independent components. In the vacuum where all components of the energy momentum tensor are zero all the components of the Riemann curvature tensor must also be zero. No action at a distance or gravity waves are allowed.
>
> Electromagnetism in Flatland, on the other hand, behaves just as one would expect. The electromagnetic field tensor in four-dimensional space–time has six independent components that can be expressed as vector E and B fields with three components each. The electromagnetic field tensor in a three-dimensional space–time (Flatland) has three independent components: a vector E field with two compo-

nents and a scalar B field. Electromagnetic radiation exists, and charges have electric fields that fall off like $1/r$.

Two more letters, published in the same issue, follow. John S. Harris, of Brigham Young University's English Department, wrote:

As I examined Alexander Keewatin Dewdney's planiversal devices in Martin Gardner's article on science and technology in a two-dimensional universe, I was struck with the similarity of the mechanisms to the lockwork of the Mauser military pistol of 1895. This remarkable automatic pistol (which had many later variants) had no pivot pins or screws in its functional parts. Its entire operation was through sliding cam surfaces and two-dimensional sockets (called hinges by Dewdney). Indeed, the lockwork of a great many firearms, particularly those of the 19th century, follows essentially planiversal principles. For examples see the cutaway drawings in *Book of Pistols and Revolvers* by W. H. B. Smith.

Gardner suggests an exhibit of machines cut from cardboard, and that is exactly how the firearms genius John Browning worked. He would sketch the parts of a gun on paper or cardboard, cut out the individual parts with scissors (he often carried a small pair in his vest pocket), and then would say to his brother Ed, "Make me a part like this." Ed would ask, "How thick, John?" John would show a dimension with his thumb and forefinger, and Ed would measure the distance with calipers and make the part. The result is that virtually every part of the 100 or so Browning designs is essentially a two-dimensional shape with an added thickness.

This planiversality of Browning designs is the reason for the obsolescence of most of them. Dewdney says in his enthusiasm for the planiverse that "such devices are invariably space-saving." They are also expensive to manufacture. The Browning designs had to be manufactured by profiling machines: cam-following vertical milling machines. In cost of manufacture such designs cannot compete with designs that can be produced by automatic screw-cutting lathes, by broaching machines, by stamping, or by investment casting. Thus although the Browning designs have a marvelous aesthetic appeal, and although they function with delightful smoothness, they have nearly all gone out of production. They simply got too expensive to make.

Stefan Drobot, a mathematician at Ohio State University, had this to say:

> In Martin Gardner's article he and the authors he quotes seem to have overlooked the following aspect of a "planiverse": any communication by means of a wave process, acoustic or electromagnetic, would in such a universe be impossible. This is a consequence of the Huygens principle, which expresses a mathematical property of the (fundamental) solutions of the wave equation. More specifically, a sharp impulse-type signal (represented by a "delta function") originating from some point is propagated in a space of three spatial dimensions in a manner essentially different from that in which it is propagated in a space of two spatial dimensions. In three-dimensional space the signal is propagated as a sharp-edged spherical wave without any trail. This property makes it possible to communicate by a wave process because two signals following each other in a short time can be distinguished.
>
> In a space with two spatial dimensions, on the other hand, the fundamental solution of the wave equation represents a wave that, although it too has a sharp edge, has a trail of theoretically infinite length. An observer at a fixed distance from the source of the signal would perceive the oncoming front (sound, light, etc.) and then would keep perceiving it, although the intensity would decrease in time. This fact would make communication by any wave process impossible because it would not allow two signals following each other to be distinguished. More practically such communication would take much more time. This letter could not be read in the planiverse, although it is (almost) two-dimensional.

My linear checkers and chess prompted many interesting letters. Abe

Schwartz assured me that on the 11-cell checker field Black also wins if the game is give-away. I. Richard Lapidus suggested modifying linear chess by interchanging knight and rook (the game is a draw), by adding more cells, by adding pawns that capture by moving forward one space, or by combinations of the three modifications. If the board is long enough, he suggested duplicating the pieces—two knights, two rooks—and adding several pawns, allowing a pawn a two-cell start option as in standard chess. Peter Stampolis proposed sustituting for the knight two pieces called "kops" because they combine features of knight and bishop moves. One kop moves only on white cells, the other only on black.

Of course many other board games lend themselves to linear forms, for example, Reversi (also called Othello), or John Conway's Phutball, described in the two-volume *Winning Ways* written by Elwyn Berlekamp, Richard Guy, and John Conway.

References

AN EPISODE OF FLATLAND. C. H. Hinton. Swan Sonnenschein & Co., 1907.

FLATLAND: A ROMANCE OF MANY DIMENSIONS, BY A. SQUARE. E. A. Abbott. Dover Publications, Inc., 1952. Several other paperback editions are currently in print.

SPHERELAND: A FANTASY ABOUT CURVED SPACES AND AN EXPANDING UNIVERSE. Thomas Y. Crowell, 1965.

THE PLANIVERSE. A. K. Dewdney. Poseidon, 1984.

ALLEGORY THROUGH THE COMPUTER CLASS: SUFISM IN DEWDNEY'S PLANIVERSE. P. J. Stewart in *Sufi*, Issue 9, pages 26–30; Spring 1991.

200 PERCENT OF NOTHING: AN EYE OPENING TOUR THROUGH THE TWISTS AND TURNS OF MATH ABUSE AND INNUMERACY. A. K. Dewdney. Wiley, 1994.

INTRODUCTORY COMPUTER SCIENCE: BITS OF THEORY, BYTES OF PRACTICE. A. K. Dewdney. Freeman, 1996.

2

Bulgarian Solitaire and Other Seemingly Endless Tasks

With useless endeavor,
Forever, forever,
Is Sisyphus rolling
His stone up the mountain!

—Henry Wadsworth Longfellow,
The Masque of Pandora

Suppose you have a basket containing 100 eggs and also a supply of egg cartons. Your task is to put all the eggs into the cartons. A step (or move) consists of putting one egg into a carton or taking one egg from a carton and returning it to the basket. Your procedure is this: After each two successive packings of an egg you move an egg from a carton back to the basket. Although this is clearly an inefficient way to pack the eggs, it is obvious that eventually all of them will get packed.

Now assume the basket can hold any finite number of eggs. The task is unbounded if you are allowed to start with as many eggs as you like. Once the initial number of eggs is specified, however, a finite upper bound is set on the number of steps needed to complete the job.

If the rules allow transferring any number of eggs back to the basket any time you like, the situation changes radically. There is no longer an upper bound on the steps needed to finish the job even if the basket initially holds as few as two eggs. Depending on the rules, the task of packing a finite number of eggs can be one that must end, one that cannot end or one that you can choose to make either finite or infinite in duration.

We now consider several entertaining mathematical tasks with the following characteristic. It seems intuitively true that you should be able to delay completing the task forever, when actually there is no way to avoid finishing it in a finite number of moves.

Our first example is from a paper by the philosopher-writer-logician Raymond M. Smullyan. Imagine you have an infinite supply of pool balls, each bearing a positive integer, and for every integer there is an infinite number of balls. You also have a box that contains a finite quantity of numbered balls. Your goal is to empty the box. Each step consists of removing a ball and replacing it with any finite number of balls of lower rank. The 1 balls are the only exceptions. Since no ball has a rank lower than 1, there are no replacements for a 1 ball.

It is easy to empty the box in a finite number of steps. Simply replace each ball higher than 1 with a 1 ball until only 1 balls remain, then take out the 1 balls one at a time. The rules allow you, however, to replace a ball with a rank above 1 with *any* finite number of balls of lower rank. For instance, you may remove a ball of rank 1,000 and replace it with a billion balls of rank 999, with 10 billion of rank 998, with a billion billion of rank 987, and so on. In this way the number of balls in the box may increase beyond imagining at each step. Can you not prolong the emptying of the box forever? Incredible as it may seem at first, there is no way to avoid completing the task.

Note that the number of steps needed to empty the box is unbounded in a much stronger way than it is in the egg game. Not only is there no bound on the number of eggs you begin with but also each time you remove a ball with a rank above 1 there is no bound to the number of balls you may use to replace it. To borrow a phrase from John Horton Conway, the procedure is "unboundedly unbounded." At every stage of the game, as long as the box contains a single ball other than a 1 ball, it is impossible to predict how many steps it will take to empty the box of all but 1 balls. (If all the balls are of rank 1, the box will of course empty in as many steps as there are 1 balls.) Nevertheless, no matter how clever you are in replacing balls, the box eventually must empty after a finite number of moves. Of course, we have to assume that although you need not be immortal, you will live long enough to finish the task.

Smullyan presents this surprising result in a paper, "Trees and Ball Games," in *Annals of the New York Academy of Sciences* (Vol. 321, pages 86-90; 1979). Several proofs are given, including a simple argument by induction. I cannot improve on Smullyan's phrasing:

> If all balls in the box are of rank 1, then we obviously have a losing game. Suppose the highest rank of any ball in the box is 2. Then we have at the outset a finite number of 2s and a finite number of 1s. We can't keep throwing away 1s forever; hence we must sooner or later throw out one of our 2s. Then we have one less 2 in the box (but possibly many more 1s than we started with). Again, we can't keep throwing out 1s forever, and so we must sooner or later throw out another 2. We see that after a finite number of steps we must throw away our last 2, and then we are back to the situation in which we have only 1s. We already know this to be a losing situation. This proves that the process must terminate if the

highest rank present is 2. Now, what if the highest rank is 3? We can't keep throwing away just balls of rank ≤ 2 forever (we just proved that!); hence we must sooner or later throw out a 3. Then again we must sooner or later throw out another 3, and so we must eventually throw out our last 3. This then reduces the problem to the preceding case when the highest rank present is 2, which we have already solved.

Smullyan also proves that the game ends by modeling it with a tree graph. By "tree" is meant a set of line segments each of which joins two points, and in such a way that every point is connected by a unique path of segments leading to a point called the tree's root. The first step of a ball game, filling the box, is modeled by representing each ball as a point, numbered like the ball and joined by a line to the tree's root. When a ball is replaced by other balls of lower rank, its number is erased and the new balls indicated by a higher level of numbered points are joined to the spot where the ball was removed. In this way the tree grows steadily upward, its "endpoints" (points that are not the root and are attached to just one segment) always representing the balls in the box at that stage of the game.

Smullyan proves that if this tree ever becomes infinite (has an infinity of points), it must have at least one infinite branch stretching upward forever. This, however, is clearly impossible because the numbers along any branch steadily decrease and therefore must eventually terminate in 1. Since the tree is finite, the game it models must end. As in the ball version, there is no way to predict how many steps are needed to complete the tree. At that stage, when the game becomes bounded, all the endpoints are labeled 1. The number of these 1 points may, of course, exceed the number of electrons in the universe, or any larger number. Nevertheless, the game is not Sisyphean. It is certain to end after a finite number of moves.

Smullyan's basic theorem, which he was the first to model as a ball game, derives from theorems involving the ordering of sets that go back to Georg Cantor's work on the transfinite ordinal numbers. It is closely related to a deep theorem about infinite sets of finite trees that was first proved by Joseph B. Kruskal and later in a simpler way by C. St. J. A. Nash-Williams. More recently Nachum Dershowitz and Zohar Manna have used similar arguments to show that certain computer programs, which involve "unboundedly unbounded" operations, must eventually come to a halt.

A special case of Smullyan's ball game is modeled by numbering a finite tree upward from the root as in Figure 11, left. We are allowed to chop off any endpoint, along with its attached segment, then add to the tree as many new branches as we like, and wherever we like, provided all the new points are of lower rank than the one removed. For example, the figure at the right in the illustration shows a possible new growth after a 4 point has been chopped off. In spite of the fact that after each chop the tree may grow billions on billions of new branches, after a finite number of chops the tree will be chopped down. Unlike the more general ball game, we cannot remove any point we like, only the endpoints, but because each removed point is replaced by points of lower rank, Smullyan's ball theorem applies. The tree may grow inconceivably bushier after each chop, but there is a sense in which it always gets closer to the ground until eventually it vanishes.

A more complicated way of chopping down a tree was proposed by Laurie Kirby and Jeff Paris in *The Bulletin of the London Mathematical Society* (Vol. 14, Part 4, No. 49, pages 285–293; July 1982). They call their tree graph a hydra. Its endpoints are the hydra's heads, and Hercules wants to destroy the monster by total decapitation. When a head is severed, its attached segment goes with it. Unfortunately after the first chop the hydra acquires one or more new heads by growing a new branch from a point (call it k) that is one step below the lost segment.

Figure 11

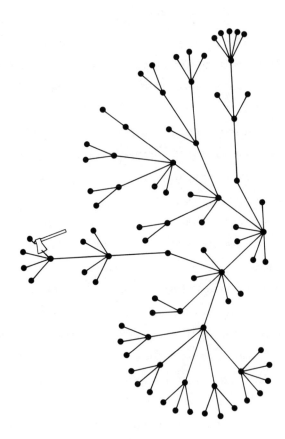

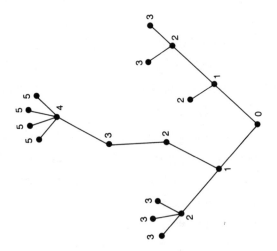

This new branch is an exact replica of the part of the hydra that extends up from k. The figure at the top right in Figure 12 shows the hydra after Hercules has chopped off the head indicated by the sword in the figure at the top left.

The situation for Hercules becomes increasingly desperate because when he makes his second chop, *two* replicas grow just below the severed segment (Figure 12, bottom left). And *three* replicas grow after the third chop (Figure 12, bottom right), and so on. In general, n replicas sprout at each nth chop. There is no way of labeling the hydra's points to make this growth correspond to Smullyan's ball game; nevertheless, Kirby and Paris are able to show, utilizing an argument based on a remarkable number theorem found by the British logician R. L. Goodstein, that no matter what sequence Hercules follows in cutting off heads, the hydra is eventually reduced to a set of heads (there

Figure 12

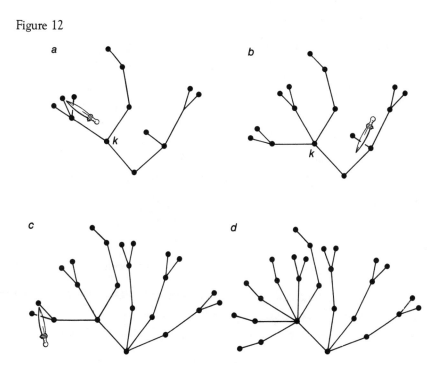

may be millions of them even if the starting form of the beast is simple) that are all joined directly to the root. They are then eliminated one by one until the hydra expires from lack of heads.

A useful way to approach the hydra game is to think of the tree as modeling a set of nested boxes. Each box contains all the boxes reached by moving upward on the tree, and it is labeled with the maximum number of levels of nesting that it contains. Thus in the first figure of the hydra the root is a box of rank 4. Immediately above it on the left is a 3 box and on the right is a 2 box, and so on. All endpoints are empty boxes of rank 0. Each time a 0 box (hydra head) is removed the box immediately below gets duplicated (along with all its contents), but each of the duplicates as well as the original box now contains one fewer empty box. Eventually you are forced to start reducing ranks of boxes, like the ranks of balls in the ball game. An inductive argument similar to Smullyan's will show that ultimately all boxes become empty, after which they are removed one at a time.

I owe this approach to Dershowitz, who pointed out that it is not even necessary for the hydra to limit its growth to a consecutively increasing number of new branches. After each chop as many finite duplicates as you like may be allowed to sprout. It may take Hercules much longer to slay the monster, but there is no way he can permanently avoid doing so if he keeps hacking away. Note that the hydra never gets taller as it widens. Some of the more complicated growth programs considered by Dershowitz and Manna graph as trees that can grow taller as well as wider, and such trees are even harder to prove terminating.

Our next example of a task that looks as if it could go on forever when it really cannot is known as the 18-point problem. You begin with a line segment. Place a point anywhere you like on it. Now place a second point so that each of the two points is within a different half of the line segment. (The halves are taken to be "closed intervals," which means that the endpoints are not considered "inside" the interval.) Place a third point so that each of the three is in a different third of the line. At this stage it becomes clear that the first two points cannot be just anywhere. They cannot, for example, be close together in the middle of the line or close together at one end. They must be carefully placed so that when the third point is added, each will be in a different third of the line. You proceed in this way, placing every nth point so that the first n points always occupy different 1/nth parts of the line. If you choose locations carefully, how many spots can you put on the line?

Intuitively it seems as if the number should be endless. A line segment obviously can be divided into as many equal parts as you like and each may contain a point. The catch is that the points must be serially numbered to meet the task's conditions. It turns out, astonishingly, that you cannot get beyond 17 points! Regardless of how clever you are at placing 17 points, the 18th will violate the rules and the game ends. In fact, it is not even easy to place 10 points. Figure 13 shows one way to place six.

This unusual problem first appeared in *One Hundred Problems in Elementary Mathematics* (problems 6 and 7) by the Polish mathematician Hugo Steinhaus. (Basic Books published a translation in 1964, and there is now a Dover soft-cover reprint.) Steinhaus gives a 14-point

Figure 13

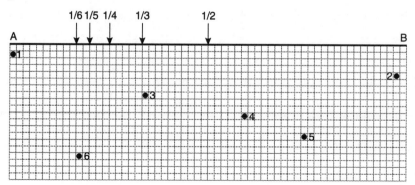

solution, and he states in a footnote that M. Warmus has proved 17 is the limit. The first published proof, by Elwyn R. Berlekamp and Ronald L. Graham, is in their paper "Irregularities in the Distributions of Finite Sequences," *Journal of Number Theory* (Vol. 2, No. 2, pages 152–161; May 1970).

Warmus, a Warsaw mathematician, did not publish his shorter proof until six years later in the same journal (Vol. 8, No. 3, pages 260–263; August 1976). He gives a 17-point solution, and he adds that there are 768 patterns for such a solution, or 1,536 if you count their reversals as being different.

Our last example of a task that ends suddenly in a counterintuitive way is one you will enjoy modeling with a deck of playing cards. Its origin is unknown, but Graham, who told me about it, says that European mathematicians call it Bulgarian solitaire for reasons he has not been able to discover. Partial sums of the series $1 + 2 + 3 + \ldots$ are known as triangular numbers because they correspond to triangular arrays such as the 10 bowling pins or the 15 pool balls. The task involves any triangular number of playing cards. The largest number you can get from a standard deck is 45, the sum of the first nine counting numbers.

Form a pile of 45 cards, then divide it into as many piles as you like, with an arbitrary number of cards in each pile. You may leave it as a single pile of 45, or cut it into two, three, or more piles, cutting anywhere you want, including 44 cuts to make 45 piles of one card each. Now keep repeating the following procedure. Take one card from each pile and place all the removed cards on the table to make a new pile. The piles need not be in a row. Just put them anywhere. Repeat the procedure to form another pile, and keep doing it.

As the structure of the piles keeps changing in irregular ways it seems unlikely you will reach a state where there will be just one pile with one card, one pile with two cards, one with three, and so on to one with nine cards. If you should reach this improbable state, without getting trapped in loops that keep returning the game to a previous state, the game must end, because now the state cannot change. Repeating the procedure leaves the cards in exactly the same consecutive state as before. It turns out, surprisingly, that regardless of the initial state of the game, you are sure to reach the consecutive state in a finite number of moves.

Bulgarian solitaire is a way of modeling some problems in partition theory that are far from trivial. The partitions of a counting number n are all the ways a positive integer can be expressed as the sum of positive integers without regard to their order. For example, the triangular number 3 has three partitions: $1 + 2$, $1 + 1 + 1$, and 3. When you divide a packet of cards into an arbitrary number of piles, any number to a pile, you are forming a partition of the packet. Bulgarian solitaire is a way of changing one partition to another by subtracting 1 from each number in the partition, then adding a number equal to the number of subtracted 1s. It is not obvious this procedure always gives rise to a chain of partitions, without duplicates, that ends with the consecutive partition. I am told it was first proved in 1981 by Jørgen Brandt, a Danish mathematician, but I do not know his proof or whether it has been published.

Bulgarian solitaire for any triangular number of cards can be diagrammed as a tree with the consecutive partition labeling its root and all other partitions represented by the tree's points. The picture at the left in Figure 14 shows the simple tree for the three-card game. In the picture at the right of the illustration is the less trivial tree for the 11 partitions of six cards. The theorem that any game ends with the consecutive partition is equivalent to the theorem that all the partitions of a triangular number will graph as a connected tree, with each partition one step above its successor in the game and the consecutive partition at the tree's root.

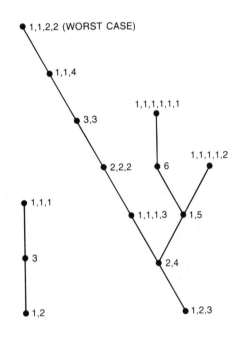

Figure 14

Note that the highest point on the six-card tree is six steps from the root. This partition, 1,1,2,2, is the "worst" starting case. It is easy to see that the game must end in no more than six steps from any starting partition. It has been conjectured that any game must end in no more than $k(k - 1)$ steps, where k is any positive integer in the formula for triangular numbers $1/2k(k + 1)$. Last year the computer scientist Donald E. Knuth asked his Stanford University students to test the conjecture by computer. They confirmed it for $k = 10$ or less, so that the conjecture is almost certainly true, but so far a proof has been elusive.

Figure 15 shows the tree for Bulgarian solitaire with 10 cards ($k = 4$). There are now three worst cases at the top, each 12 steps from the root. Note also that the tree has 14 endpoints. We can call them Eden partitions because unless you start with them they never arise in a game. They are all those partitions whose number of parts exceeds the highest number of parts by 2 or more.

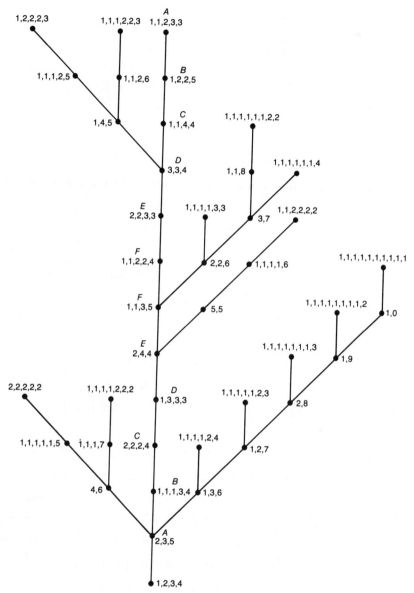

Figure 15

The picture at the left in Figure 16 shows the standard way of using dots to diagram partition 1,1,2,3,3, at the tree's top. If this pattern is rotated and mirror-reflected, it becomes the pattern at the right in the

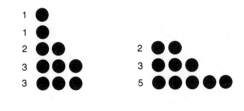

Figure 16

illustration. Its rows now give the partition 2,3,5. Each partition is called the conjugate of the other. The relation is obviously symmetrical. A partition unchanged by conjugation is said to be self-conjugate. On the 10 tree there are just two such partitions, the root and 1,1,1,2,5. When the remaining partitions are paired as conjugates, an amazing pattern appears along the trunk. The partitions pair as is shown by the letters. This symmetry holds along the main trunk of all Bulgarian trees so far investigated.

If the symmetry holds for all such trees, we have a simple way to determine the worst case at the top. It is the conjugate of the partition (there is always only one) just above the root. An even faster way to find the trunk's top is to prefix 1 to the root and diminish its last number by 1.

The Bulgarian operation can be diagrammed by removing the leftmost column of its flush-left dot pattern, turning the column 90 degrees, and adding it as a new row. Only diagrams of the 1, 2, 3, 4. . . form are unaltered by this. If you could show that no sequence of operations on any partition other than the consecutive one would return a diagram to its original state, you would have proved that all Bulgarian games graph as trees and therefore must end when their root is reached.

If the game is played with 55 cards ($k = 10$), there are 451,276 ways to partition them, so that drawing a tree would be difficult. Even the 15-card tree, with 176 points, calls for computer aid. How are these numbers calculated? Well, it is a long and fascinating story. Let us say partitions are *ordered*, so that 3, for example, would have four ordered partitions (usually called "compositions"): $1 + 2$, $2 + 1$, $1 + 1 + 1$, and 3. It turns out that the formula for the total number of compositions is simply 2^{n-1}. But when the partitions are unordered, as they are in the solitaire card game, the situation is unbelievably disheveled. Although there are many recursive procedures for counting unordered partitions, using at each step the number of known partitions for all smaller numbers, an exact asymptotic formula was not obtained until recent times. The big breakthrough was made by the British mathematician G. H. Hardy, working with his Indian friend Srinivasa Ramanujan. Their not quite exact formula was perfected by Hans A. Rademacher in 1937. The Hardy–Ramanujan–Rademacher formula is a horribly shaggy infinite series that involves (among other things) pi, square roots, complex roots, and derivatives of hyperbolic functions! George E. Andrews, in his standard textbook on partition theory, calls it an "unbelievable identity" and "one of the crowning achievements" in the history of his subject.

The sequence of partitions for $n = 1$, $n = 2$, $n = 3$, $n = 4$, $n = 5$, and $n = 6$ is 1,2,3,5,7,11, and so you might expect the next partition to be the next prime, 13. Alas, it is 15. Maybe all partitions are odd. No, the next partition is 22. One of the deep unsolved problems in partition theory is whether, as n increases, the even and odd partitions approach equality in number.

If you think partition theory is little more than a mathematical pastime, let me close by saying that a way of diagramming sets of partitions, using number arrays known as the Young tableaux, has become enormously useful in particle physics. But that's another ball game.

ADDENDUM

Many readers send proofs of the conjecture that Bulgarian solitaire must end in k $(k - 1)$ steps, and the proof was later given in several articles listed in the bibliography. Ethan Akin and Morton Davis began their 1983 paper as follows:

Blast Martin Gardner! There you are, minding your own business, and *Scientific American* comes along like a virus. All else forgotten, you must struggle with infection by one of his fascinating problems. In the August 1983 issue he introduced us to Bulgarian solitaire.

References

ON WELL-QUASI-ORDERING FINITE TREES. C. Si. J. A. Nash-Williams in *Proceedings of the Cambridge Philosophical Society*, Vol. 59, Part 4, pages 833–835; October 1963.

NUMBER THEORY: THE THEORY OF PARTITIONS. George E. Andrews. Addison-Wesley Publishing Co., 1976.

TREES AND BALL GAMES. Raymond M. Smullyan in *Annals of the New York Academy of Sciences*, Vol. 321, pages 86–90; 1979.

PROVING TERMINATION WITH MULTISET ORDERINGS. Nachum Dershowitz and Zohar Manna in *Communications of the ACM*, Vol. 22, No. 8, pages 465–476; August 1979.

ACCESSIBLE INDEPENDENCE RESULTS FOR PEANO ARITHMETIC. Laurie Kirby and Jeff Paris in *The Bulletin of the London Mathematical Society*, Vol. 14, No. 49, Part 4, pages 285–293; July 1983.

CYCLES OF PARTITIONS. Jørgen Brandt in *Proceedings of the American Mathematical Society*, Vol. 85, pages 483–486; July 1982.

BULGARIAN SOLITAIRE. Ethan Akin and Morton Davis in the *American Mathematical Monthly*, Vol. 92, pages 237–250; April 1985.

SOLUTION OF THE BULGARIAN SOLITAIRE CONJECTURE. Kiyoshi Igusa in *Mathematics Magazine*, Vol. 58, pages 259–271; November 1985.

HERCULES HAMMERS HYDRA HERD. Maxwell Carver in *Discover*, pages 94–95, 104; November 1987.

SOME VARIANTS OF FERRIER DIAGRAMS. James Propp in the *Journal of Combinatorial Theory*, Series A, Vol. 52, pages 98–128; September 1989.

THE TENNIS BALL PARADOX. R. W. Hamming in *Mathematics Magazine*, Vol. 62, pages 268–273; October 1989.

BULGARIAN SOLITAIRE. Thomas Bending in *Eureka*, No. 50, pages 12–19; April 1990.

TABLEAUX DE YOUNG ET SOLITAIRE BULGARE. Gwihen Etienne in *The Journal of Combinatorial Theory*, Series A, Vol. 58, pages 181–197; November 1991.

BULGARIAN SOLITAIRE. Al Nicholson in *Mathematics Teacher*, Vol. 86, pages 84–86; January 1993.

3

Fun
with
Eggs,
Part I

Not quite
spherical
White
Oddly closed
And without a lid

—May Swenson

Thus begins "At Breakfast," eight whimsical stanzas about cracking and eating a soft-boiled egg in the Continental manner. The poem continues: "A smooth miracle/here in my hand/has it slid/from my sleeve?/ The shape/of this box/keels me oval."

Is there any natural and simple sculpture that pleases the eye and the hand more than a chicken egg? One end of the object is more pointed than the other, and the delightful oval shape varies widely from egg to egg. The shape of a chicken egg can be simulated mathematically by a host of closed curves with different low-degree formulas. The simplest curve is the oval of Descartes, a family of egg-shaped ovals discovered by the 17th-century French mathematician and philosopher. Just as an ellipse can be constructed easily with two pins and a piece of thread, so can certain Cartesian ovals.

Figure 17 shows how an ellipse is drawn by keeping taut a triangular loop of thread (nylon is best because it minimizes friction) as a pencil point traces the curve. Because the sum of AP and BP in the illustration cannot vary, the method ensures that the curve is the locus

Figure 17

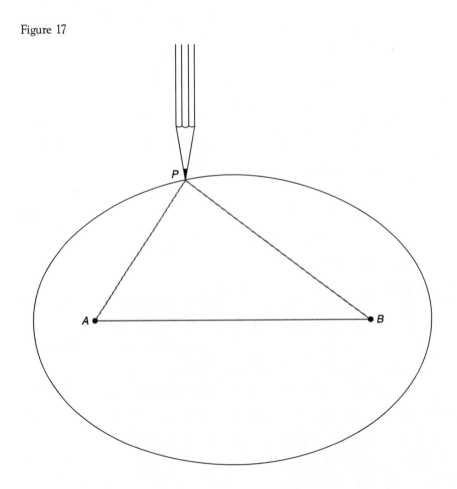

of all points the sum of whose distances from the two foci A and B is a constant.

Figure 18 shows how a Cartesian oval can be generated by a similar method. Here the thread is looped once around the pin at B and attached to the pencil point. By keeping the thread taut the upper half of the oval can be drawn. The lower half of the oval can be constructed by the same procedure with the thread arrangement inverted.

Figure 18

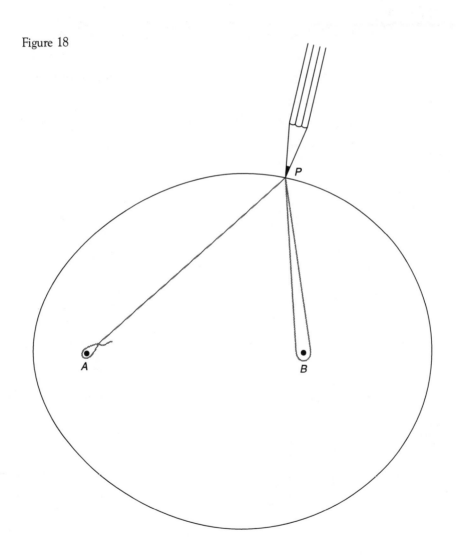

This method obviously generates a curve that is the locus of all points such that their distance from A added to twice their distance from B is a constant. Descartes generalized the curve by letting the constant be the sum of m times the distance from A and n times the distance from B, where m and n are real numbers. The ellipse and the circle are special cases of Cartesian ovals. In the ellipse m equals n, and n equals 1. The circle is an ellipse in which the distance between the foci is zero.

In the oval in the illustration, m equals 1 and n equals 2. By varying the distance between the foci, by changing the length of the thread or by doing both, it is possible to draw an infinite number of Cartesian ovals all with multipliers in the ratio 1 : 2. Figure 19 shows how to construct a family of Cartesian ovals with multipliers in the ratio 2 : 3. Here one focus lies outside the oval. Of course, the

Figure 19

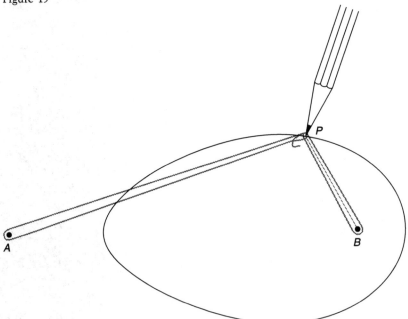

thread technique works only if m and n are positive integers and are small enough to ensure that the looping of the thread does not generate too much friction.

Many eminent physicists, including Christian Huygens, James Clerk Maxwell, and Isaac Newton, were fascinated by Cartesian ovals because of their unusual optical properties of reflection and refraction. In 1846 the Royal Society of Edinburgh heard Maxwell's paper "On the Description of Oval Curves and Those Having a Plurality of Foci." The Scottish physicist had independently discovered the ovals of Descartes. He went further, however, in generalizing them to curves with more than two foci. Maxwell did not present the paper to the society himself because, being just 15, he was considered too young to appear before such a distinguished audience! (Young Maxwell's paper is included in the Dover reprint *The Scientific Papers of James Clerk Maxwell.*)

Among the many other ovals that resemble eggs, more rounded at one end than the other, are the well-known ovals of Cassini. A Cassini oval is the locus of all points the *product* of whose distances from two fixed points is a constant. Not all Cassini ovals are egg-shaped, but when they are, they come in pairs that point in opposite directions.

The physical properties of chicken eggs make possible a variety of entertaining parlor tricks. If you try the following eggsperiments, you will find them both amusing and scientifically instructive.

Surely the oldest of all tricks with eggs is making a raw egg stand on end. Christopher Columbus is said to have done it by setting the egg down firmly enough to crush its bottom end slightly. A neater solution is to put a small quantity of salt on a white table top, balance the egg on the salt, and then gently blow away all but the few invisible salt grains that keep the egg upright. (For details about Piet Hein's supereggs, solid forms that balance on end without any skulduggery, see Chapter 18 of my *Mathematical Carnival.*)

In fact, on an unsmooth surface such as a sidewalk or a tablecloth it is not difficult to balance a raw chicken egg on its broad end with the aid of only patience and a steady hand. Now and then the practice becomes a local mania. For example, the April 9, 1945, issue of *Life* described an egg-balancing craze that had hit Chungking. According to a folk belief in China, eggs balance more easily on Li Chun, the first day of spring in the Chinese calendar.

In Figure 20 there is shown a marvelous old egg-balancing stunt with a cork, a bottle, and two forks. Hollow out one end of the cork so that it fits snugly on the egg. The forks should be long ones with heavy handles, and the rim of the bottle must be flat like that of most soft-drink bottles. Even so, it may take many minutes to make a stable structure. Because the egg may fall a few times before you balance it try a hard-boiled egg rather than a raw one. Once the precarious balance is

Figure 20

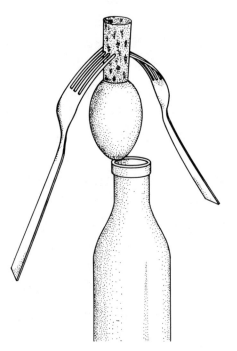

achieved it will seem mysterious to anyone who is not familiar with physical laws about the center of gravity.

Egg balancing is the secret of winning an old puzzle game. The game calls for a large supply of nearly identical eggs. Two players take turns putting an egg on a circular or square table. The loser is the player who is unable to put down an egg without moving another one. The first player can always win by standing the first egg on its end at the center of the table. On his subsequent turns he puts an egg symmetrically opposite wherever his opponent puts one.

Because the inside of a raw egg is viscous the inertial drag of the liquid makes it difficult to spin the egg on its side and impossible to spin it on its end. This provides a quick way of distinguishing a raw egg from a hard-boiled one: only a hard-boiled egg can be spun on its end. The following stunt with a raw egg is less familiar. Spin the egg on its side as fast as you can, then make it come to a dead stop by pressing it with a fingertip. Quickly remove your finger. The inertia of the rotating interior will start the egg slowly turning again.

Charlie Miller, a magician friend of mine, likes to do a surprising trick with a hard-boiled egg. He explains that the egg can spin on its side (he spins it gently on its side) and also can spin on its end (he demonstrates that) but that only a magician can make it undergo both kinds of rotation in the course of a single maneuver. At that point he spins it vigorously on its side. Most eggs (particularly ones that were kept upright during boiling) will rotate for a while and then suddenly assume a vertical spinning position. (You will find this explained in the Dover reprint *Spinning Tops and Gyroscopic Motion: A Popular Exposition of Dynamics of Rotation*, by John Perry, and in "The Amateur Scientist," by Jearl Walker, *Scientific American*, October 1979).

The most remarkable of all egg-spinning tricks is hardly ever done, probably because it takes much practice and is easier to learn from someone who can do it than from printed instructions. You will need a dinner plate with a flat rim. From the shell of an opened egg break off a piece roughly the size of a half-dollar. It will be ragged at the edge and it should come from the egg's side, not its end.

Dip the plate in water, put the piece of shell on the edge of the flat rim, and tip the plate at the angle shown in Figure 21. The shell should start to rotate. If you now turn the plate in your hands while keeping it at the same angle, the shell will spin with surprising rapidity as it travels precariously around the wet rim. To get it right you may have to try different pieces of shell until you find one with the proper balance and convexity. Once you acquire the knack you will easily be able to demonstrate this amazing juggling feat whenever you want. Although the trick is described in old conjuring books, few magicians seem to know about it.

Inertia is the secret behind the following bet. Get a kitchen knife with a sharp point, hold it vertically, and hang half an eggshell over the

Figure 21

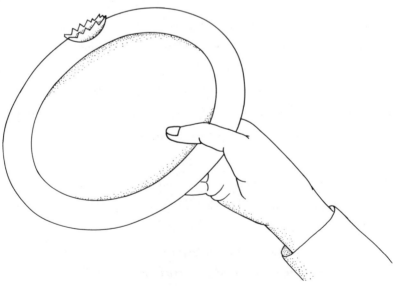

point as shown in Figure 22. Give the knife to someone and challenge him to puncture the shell by rapping the handle on a table or kitchen counter. Each time he tries the shell will bounce off unharmed, whereas you can crack it at will. The secret is to hold the blade loosely in your hand. Make it look as if you rap the handle on the counter when in fact

Figure 22

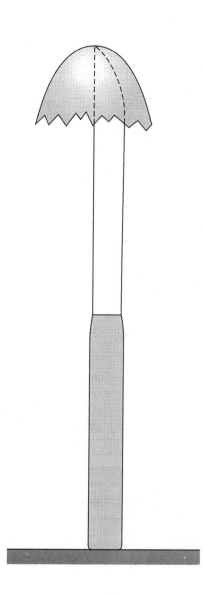

you allow the knife to fall by its own weight so that it hits the counter and bounces. The imperceptible bounce sends the knife point through the shell.

The intact shell surrounding a raw egg is remarkably strong. Many people know that if you clasp your hands with an egg between them, each end touching the center of a palm, it is almost impossible to break the egg by squeezing. What is not so well known is the difficulty of smashing a raw egg by tossing it high into the air and letting it fall onto grass. The May 18, 1970, issue of *Time* described a flurry of such experiments that took place at Richmond in England after the headmaster of a school did it for his students. A local fireman dropped raw eggs onto grass from the top of a 70-foot ladder. Seven out of 10 survived. An officer in the Royal Air Force arranged for a helicopter to drop eggs from 150 feet onto the school's lawn. Only three out of 18 broke. *The Daily Express* hired a Piper Aztec to dive-bomb an airfield with five dozen eggs at 150 miles per hour. Three dozen of them were unharmed. When eggs were dropped into the Thames from Richmond Bridge, three-fourths of them shattered. That proved, said the school's science teacher, "that water is harder than grass but less hard than concrete."

The fragility of an egg when it falls onto a hard surface is the subject of the old nursery rhyme about Humpty-Dumpty and its retelling by Lewis Carroll in *Though the Looking-Glass*. It is also involved in the following practical joke. Bet someone a dime that he cannot put a thumb and finger through the crack of a door above the top hinge and hold a raw egg for 30 seconds on the other side of the crack. As soon as he firmly grasps the egg, put his hat on the floor directly below the egg, walk away, and forget about him.

The best of all scientific tricks with an egg is the well-known one in which air pressure forces a peeled hard-boiled egg into a glass milk bottle and then forces it out again undamaged. The mouth of the bottle must be only slightly smaller than the egg, and so you must be careful not to use too large an egg or too small a bottle. It is impossible to push the egg into the bottle. To get the egg through the mouth you must heat the air in the bottle. That is best done by standing the bottle in boiling water for a few minutes. Put the egg upright on the mouth and take the bottle off the stove. As the air in the bottle cools it contracts, creating a partial vacuum that allows air pressure to push the peeled egg inside. To get the egg out again invert the bottle so that the egg falls into the neck. Place the opening of the bottle against your mouth and blow vigorously. This will compress the air in the bottle. When you stop blowing, the air expands, pushing the egg through the neck of the bottle and into your waiting hand.

Many old books suggest the following elaboration using a hardboiled egg with its shell in place. Soak the egg for a few hours in heated vinegar until the shell becomes pliable. Put the egg into a bottle by the method described above and let it soak overnight in cold water. The shell will harden. Pour out the water and you have a curiosity with which to puzzle friends. It happens, however, that I have never been able to make this work. The shell does soften, but it also seems to become porous, which prevents a vacuum from forming. (I would be interested in hearing from any reader who can tell me how to get the vinegar-treated egg into the bottle.) Regardless of whether or not the feat works, the failure to perform it is central to one of Sherwood Anderson's funniest and finest short stories. It is called "The Egg." You will find it in his book *The Triumph of the Egg.*

The story is told by a boy. His parents, who formerly owned a miserable chicken farm, have bought a restaurant across the road from the railway station at Pickleville, a place not far from Bidwell, Ohio. The father fancies himself a showman. One rainy night the only customer in the restaurant is Joe Kane, a young man who is waiting for a late train. The father decides to amuse him by performing his favorite egg trick.

"I will heat this egg in this pan of vinegar," he says to Joe. "Then I will put it through the neck of a bottle without breaking the shell. When the egg is inside the bottle, it will resume its normal shape and the shell will become hard again. Then I will give the bottle with the egg in it to you. You can take it about with you wherever you go. People will want to know how you got the egg in the bottle. Don't tell them. Keep them guessing. That is the way to have fun with this trick."

When the father grins and winks, Joe decides the man is crazy but harmless. The vinegar softens the egg's shell but the father forgets an essential part of the trick. He neglects to heat the bottle.

"For a long time he struggled, trying to get the egg to go through the neck of the bottle. . . . He worked and worked and a spirit of desperate determination took possession of him. When he thought that at last the trick was about to be consummated, the delayed train came in at the station and Joe Kane started to go nonchalantly out at the door. Father made a last desperate effort to conquer the egg and make it do the thing that would establish his reputation as one who knew how to entertain guests who came into his restaurant. He worried the egg. He attempted to be somewhat rough with it. He swore and the sweat stood out on his forehead. The egg broke under his hand. When the contents spurted over his clothes, Joe Kane, who had stopped at the door, turned and laughed."

Roaring with anger, the father grabs another egg and hurls it at Joe, just missing him. Then he closes the restaurant for the night and tramps upstairs, where his wife and son have been awakened by the noise. There is an egg in his hand and an insane gleam in his eyes. He gently puts the egg on the table by the bed and begins to cry. The boy, caught up in his father's grief, weeps with him.

Good stories have a way of turning into allegories. What does the egg represent? I think it is nature, the Orphic Egg, the vast world that is independent of our minds, under no obligation to conform to our desires. Understand its mathematical laws and you can control it to an incredible degree, as modern science and technology testify. Fail to understand its laws or forget them or ignore them and nature can be as malevolent as Moby Dick, the white whale, or the white egg in Anderson's tragedy.

An egg is an egg is an egg. It is a small physical thing with a beautiful geometrical surface. It is a microcosm that obeys all the laws of the universe. And at the same time it is something far more complex and mysterious than a white pebble. It is a strange lidless box that holds the secret of life itself. As May Swenson continues in her poem:

> Neatly
> The knife scalps it
> I scoop out
> the braincap
> Soft
> Sweetly shuddering

Which is more important, the chicken or the egg? Is the hen, as Samuel Butler said, no more than an egg's way of making another egg? Or is it the other way around?

"I awoke at dawn," Anderson's narrator concludes his account of human failure, "and for a long time looked at the egg that lay on the table. I wondered why eggs had to be and why from the egg came the hen who again laid the egg. The question got into my blood. It has stayed there, I imagine, because I am the son of my father. At any rate, the problem remains unsolved in my mind. And that, I conclude, is but another evidence of the complete and final triumph of the egg—at least as far as my family is concerned."

References

A BOOK OF CURVES. Edward H. Lockwood. Cambridge University Press, 1961.

MAXWELL'S OVALS AND THE REFRACTION OF LIGHT. Milton H. Sussman in *American Journal of Physics*, Vol. 34, No. 5, pages 416–418; May 1966.

EGGS. Martin Gardner in *The Encyclopedia of Impromptu Magic*. Chicago: Magic, 1978.

THE DRAWING-OUT OF AN EGG. Robert Dixon in *New Scientist*, pages 290–295; July 29, 1982.

EGGS: NATURE'S PERFECT PACKAGE. Robert Burton. Facts on File, 1987.

4

Fun with Eggs, Part II

So much accumulated in my files about eggs that I decided it deserved a new chapter rather than a lengthy addendum to the previous one.

Glass milk bottles are hard to find these days. The best kind of bottle to use for the trick of putting an egg inside is a bottle used as a wine carafe.

To heat the air inside the bottle, it is often recommended that a burning piece of paper or a short portion of a lit candle be dropped into the bottle. The instructions frequently add that the vacuum in the bottle is created by the burning of oxygen. Not so. The burning of oxygen plays no role. The vacuum is caused entirely by the cooling and contraction of air.

Dozens of readers offered suggestions for getting the vinegar-treated egg into the bottle, though most of them did not actually try to do it. Many thought it would help to coat the egg with some sort of sealant such as oil, syrup, honey, or vaseline. Several proposed sealing the mouth of the bottle by placing the egg on top, then wrapping it with kitchen plastic wrap around both the egg and the bottle's neck. Several readers said the best way to heat the air is to fill the bottle halfway with water, then boil the water. As reader Kevin Miller explained, condensing water vapor creates a stronger vacuum than cooling air. The water also serves to cushion the shock when the egg pops down.

Before placing the egg on top, Miller squeezed toothpaste around the bottle's rim. Running cold water on the bottle's side created the vacuum. However, after the egg had soaked in water for two months the shell never hardened. One reader thought soaking the egg in a borax solution would do the job by neutralizing the acidic egg. Another said the shell would harden if the egg were thoroughly rinsed with cold running water. I myself have been unable to get the shell to return to normal. Perhaps this is a myth created by old books on recreational science?

Lakenan Barnes, an attorney in Mexico, Missouri, sent the following letter:

> Your article on the fun with eggs in the April issue of *Scientific American* was eggscellent and eggscited me. I am never sure what you are going to cook up next, which, of course, pleases me.
>
> After hard cooking an egg as you directed and spinning it, I was eggstremely flattered when it gave me several stand-

ing ova-tions. Since its shell is calcium carbonate, I consider it a marble to behold.

I have demonstrated this lesson in physics to many of my coffee counter buddies on several occasions, but I use a decorative alabaster egg, obtaining the same results but without the danger of someone, Pilate-like, proclaiming, "Eggy homo!"

Mr. Gardner, I hope my poor yolks do not eggsasperate you, but I want you to know how delighted I was with your discussion of your "eggsperiment."

One last thought: As for the chicken-or-egg dilemma, I agree with my Amherst fraternity brother who, over fifty years ago, remarked, "Of course the egg came first!" I accept this without proof even though I am still from Missouri.

In April 1990 the cartoon strip "The Wizard of Id" raised a curious question. "Did you ever consider how brave the first man to eat an egg was?" asked a parrot. "I never though of it before, but you're right," says the man listening to the parrot. To which the parrot replies, "What's even more astonishing is that it caught on!"

Pendleton Tompkins, a physician in San Mateo, California, wrote as follows:

I am told that if a small foreign body is placed in the peritoneal cavity of a laying hen, she will encase it in a shell and lay it. A Professor of Animal Husbandry made use of this phenomenon when he was entertaining a dozen colleagues at breakfast. He prepared a number of small slips of paper upon which was written "Hello Bill" or "Good Morning Joe" and placed each in a capsule. He then put the capsules (which were radiopaque) in a dozen laying hens who were labeled Bill, Joe, and so forth. As eggs were laid he put each under a fluoroscope until he discovered the one containing the capsule. Each egg was labeled Bill, Joe, etc., and was served as an oeuf à le coq at breakfast. Imagine the surprise of the guests to open an egg and find therein a capsule greeting them by name.

The notion that fresh eggs balance more easily on the day of the vernal equinox turned into a minor craze in the United States. For a history of this bizarre phenomenon, and an explanation of why so many seemingly intelligent people take such egg balancing seriously, see my column "Notes of a Fringe Watcher" in *The Skeptical Inquirer* (May/June 1996). The column also discusses a variety of mechanical eggs that stand on their broad ends only if you know the secret of how to make them balance.

I ended the foregoing chapter with the old riddle of which came first, the chicken or the egg? The answer is the egg. Like all birds, chickens evolved from reptiles. Because reptiles lay eggs, eggs preceded chickens; but which came first, the reptile or the egg?

May Swenson, whose poetry I admire (one of her poems provided the previous chapter's epigraph), was born in Logan, Utah, in 1919, and died in 1989. She had a great love of puzzles and word play, displayed most notably in her collections *Poems to Solve* (1966) and *More Poems to Solve* (1971).

Mary J. Packard, then a research associate in the zoology department of Colorado State University, Fort Collins, sent me several technical papers about eggs on which she had collaborated. In her letter she praised eggs for being "easy to house, have minimal nutritional requirements, do not bite, run slowly at best, and are eggceedingly easy to trap." To reward me for my efforts to educate readers about "the perfection of eggs," she bestowed on me a Good Egg reward.

More seriously, Ms. Packard raised the interesting question of why the air cell in all bird eggs is always at the blunt end. This cell is important because the embryo uses it for the first inflation of its lungs, and can die if the cell is not there. It is convenient, she adds, that the air cell always be at a predictable spot, but why it forms only at the blunt end, not at the side or the pointed end, seems to be an unsolved zoological mystery.

The air cells, by the way, are the basis for an amusing bet. Here is how I described it in *Physics Teacher*, as one of my "Physics Trick of the Month" features:

> Carefully open a fresh egg so that the two half-shells are as similar as possible. Check to make sure that the shell for the larger end has an air bubble inside. Most eggs do.
>
> If the bubble is there, you can mystify a friend with the following stunt. Fill a tall glass with water. Give the shell without the air bubble to the friend, while you keep the other half-shell. Say nothing about the bubble.
>
> Put your half-shell, open side up, on top of the water, and gently push on the shell's rim until the shell fills with water and submerges. As it sinks, the bubble will cause it to flip over and land convex end up. Fish it out with a spoon and challenge your victim to duplicate the feat. When he tries, his shell stubbornly refuses to turn over.
>
> Repeat a few times. After the last somersault, surreptitiously poke your finger into the shell to break the bubble. If your friend thinks your shell differs in some way from his, let him now try it with your shell. To his puzzlement, the shell *still* refuses to flip over.

Two readers, Frank Colon and Fred Kolm, independently sent the following method of drawing an egg by using a closed loop of string and *three* pins. Figure 23 shows how this is done. The oval consists of six segments of ellipses, all interrelated and tangent to each other. I sent the two letters to Professor H. S. M. Coxeter, the University of Toronto's famous geometer. He replied that perhaps the reason the three-pin method is so little known is because its oval is an artificial composite of elliptical segments with no simple generating equation. His own favorite, he added, is the cubic $y^2 = (x - a)(x - b)(x - c)$, ignoring the curve's infinite branch.

Some more odds and ends:

Sherwood Anderson wrote a sort of companion to his "Triumph of

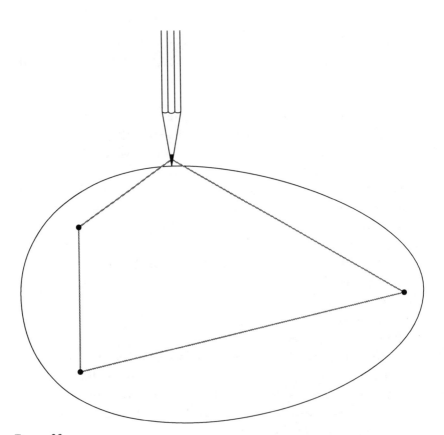

Figure 23

the Egg," titled "Milk Bottles." You'll find the story in his collection
Horses and Men.

Paul Outerbridge, Jr., in 1932, painted a striking picture titled "The
Triumph of the Egg." I would have reproduced it here except that the
Los Angeles gallery exhibiting his work wanted $125 for reproduction
rights.

Old riddle: What did the hen say when she laid a cubical egg?
"Ouch!" A Paul Bunyan tall tale, repeated in Jorge Luis Borges' book
Imaginary Beasts, tells of the Gillygaloo bird that lays cubical eggs to
keep them from rolling down hills. Lumberjacks hardboil the eggs to
use for dice.

In *Gulliver's Travels* Swift describes a fierce war fought over the proper way to crack open an egg. He was, of course, satirizing religious wars fought over how to interpret a theological doctrine. The Nomes, in L. Frank Baum's Oz books, live in underground caves in Ev, a magic land adjacent to Oz. They are deathly afraid of eggs because if they touch one they lose their immortality and become subject to old age and death. If they touch the inside of an egg they wither up and blow away.

Sextus Empiricus, the ancient Greek skeptic from whom the word "empiricism" derives, had the following to say in his book *Against the Logicians*, Vol. 2: "Others say that philosophy resembles an egg, ethics being like the yolk, which some identify with the chick, physics like the white, which is nutrient for the yolk, and logic like the outside shell."

George Santayana, in the last chapter of *Dialogues in Limbo*, provides a satirical explanation of Aristotle's four kinds of causes as they relate to eggs: the efficient cause is the warmth of the hen, the egg's essence is the formal cause, the final cause is the chicken to be hatched, and the material cause is "a particular yolk and a particular shell and a particular farmyard, on which and in which the other three causes may work, and laboriously hatch an individual chicken, probably lame and ridiculous despite so many sponsors."

5

The
Topology
of
Knots

"A knot!" said Alice, always ready
to make herself useful, and looking
anxiously about her. "Oh, do let me
help to undo it!"

—*Alice in Wonderland*, Chapter 3

To a topologist knots are closed curves embedded in three-dimensional space. It is useful to model them with rope or cord and to diagram them as projections on a plane. If it is possible to manipulate a closed curve—of course, it must not be allowed to pass through itself—so that it can be projected on a plane as a curve with no crossing points, then the knot is called trivial. In ordinary discourse one would say the curve is not knotted. "Links" are two or more closed curves that cannot be separated without passing one through another.

The study of knots and links is now a flourishing branch of topology that interlocks with algebra, geometry, group theory, matrix theory, number theory, and other branches of mathematics. Some idea of its depth and richness can be had from reading Lee Neuwirth's excellent article "The Theory of Knots" in *Scientific American* (June 1979). Here we shall be concerned only with some recreational aspects of knot theory: puzzles and curiosities that to be understood require no more than the most elementary knowledge of the topic.

Let us begin with a question that is trivial but that can catch even mathematicians off guard. Tie an overhand knot in a piece of rope as is shown in Figure 24. If you think of the ends of the rope as being joined, you have tied what knot theorists call a trefoil knot. It is the

Figure 24

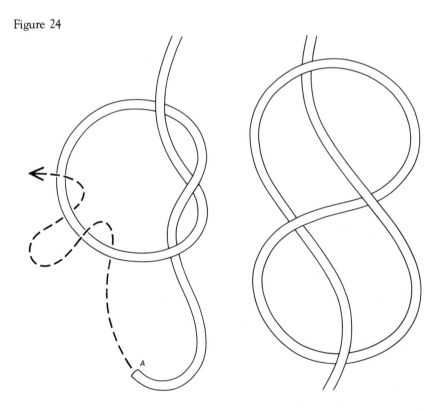

simplest of all knots in the sense that it can be diagrammed with a minimum of three crossings. (No knot can have fewer crossings except the trivial knot that has none.) Imagine that end A of the rope is passed through the loop from behind and the ends are pulled. Obviously the knot will dissolve. Now suppose the end is passed *twice* through the loop as is indicated by the broken line. Will the knot dissolve when the ends of the rope are pulled?

Most people guess that it will form another knot. Actually the knot dissolves as before. The end must go *three* times through the loop to produce another knot. If you try it, you will see that the new trefoil created in this way is not the same as the original. It is a mirror image. The trefoil is the simplest knot that cannot be changed to its mirror image by manipulating the rope.

The next simplest knot, the only one with a minimum of four crossings, is the figure eight at the right in Figure 24. In this form it is easily changed to its mirror image. Just turn it over. A knot that can be manipulated to make its mirror image is called amphicheiral because like a rubber glove it can be made to display either handedness. After the figure eight the next highest amphicheiral knot has six crossings, and it is the only 6-knot of that type. Amphicheiral knots become progressively scarcer as crossing numbers increase.

A second important way to divide knots into two classes is to distinguish between alternating and nonalternating knots. An alternating knot is one that can be diagrammed so that if you follow its curve in either direction, you alternately go over and under at the crossings. Alternating knots have many remarkable properties not possessed by nonalternating knots.

Still another important division is into prime and composite knots. A prime knot is one that cannot be manipulated to make two or more separated knots. For example, the square knot and the granny knot are not prime because each can be changed to two side-by-side trefoils. The square knot is the "product" of two trefoils of opposite handedness. The granny is the product of two trefoils of the same handedness, and therefore (unlike the square knot) it is not amphicheiral. Both knots are alternating. As an easy exercise, see if you can sketch a square knot with six (the minimum) alternating crossings.

All prime knots of seven or fewer crossings are alternating. Among the 8-knots only the three in Figure 25 are nonalternating. No matter how long you manipulate a rope model of one of these knots, you will never get it to lie flat in the form of an alternating diagram. The knot at top right is a bowline. The bottom knot is a torus knot as explained below.

A fourth basic binary division of knots is into the invertible and

Figure 25

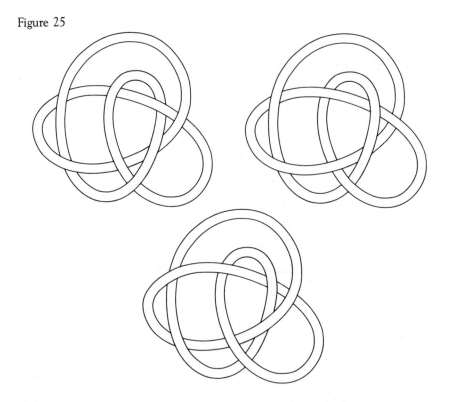

noninvertible. Imagine an arrow painted on a knotted rope to give a direction to the curve. If it is possible to manipulate the rope so that the structure remains the same but the arrow points the other way, the knot is invertible. Until the mid-1960s one of the most vexing unsolved problems in knot theory was whether noninvertible knots exist. All knots of seven or fewer crossings had earlier been found invertible by manipulating rope models, and all but one 8-knot and four 9-knots. It was in 1963 that Hale F. Trotter, now at Princeton University, announced in the title of a surprising paper "Non-invertible Knots Exist" (*Topology*, Vol. 2, No. 4, pages 275–280; December 1963.)

Trotter described an infinite family of pretzel knots that will not invert. A pretzel knot is one that can be drawn, without any crossings, on the surface of a pretzel (a two-hole torus). It can be drawn as shown in Figure 26 as a two-strand braid that goes around two "holes," or it can be modeled by the edge of a sheet of paper with three twisted strips. If the braid surrounds just one hole, it is called a torus knot because it can be drawn without crossings on the surface of a doughnut.

Trotter found an elegant proof that all pretzel knots are noninvertible if the crossing numbers for the three twisted strips are distinct odd integers with absolute values greater than 1. Positive integers indicate braids that twist one way and negative integers indicate an opposite twist. Later Trotter's student Richard L. Parris showed in his unpublished Ph.D. thesis that the absolute values can be ignored provided the signed values are distinct, and that these conditions are necessary as well as sufficient for noninvertible pretzels. Thus the simplest noninvertible pretzel is the one shown. Its crossing numbers of 3, −3, and 5 make it an 11-knot.

It is now known that the simplest noninvertible knot is the amphicheiral 8-knot in Figure 27. It was first proved noninvertible by Akio Kawauchi in *Proceedings of the Japan Academy* (Vol. 55, Series A, No. 10, pages 399–402; December 1979). According to Richard Hartley,

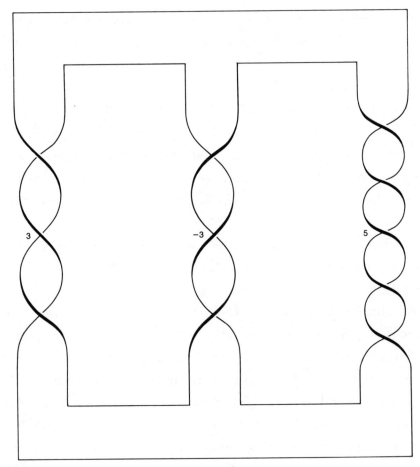

Figure 26

Figure 27

in "Identifying Non-invertible Knots" (*Topology*, Vol. 22, No. 2, pages 137–145; 1983), this is the only noninvertible knot of eight crossings, and there are only two such knots of nine crossings and 33 of 10. All 36 of these knots had earlier been declared noninvertible by John Horton Conway, but only on the empirical grounds that he had not been able to invert them. The noninvertible knots among the more than 550 knots with 11 crossings had not yet been identified.

In 1967 Conway published the first classification of all prime knots with 11 or fewer crossings. (A few minor errors were corrected in a later printing.) You will find clear diagrams for all prime knots through ten crossings, and all links through nine crossings, in Dale Rolfsen's valuable 1990 book *Knots and Links*. There are no knots with 1 or 2 crossings, one with 3, one with 4, two with 5, three with 6, seven with 7, 21 with 8 crossings, 49 with 9, 165 with 10, and 552 with 11, for a total of 801 prime knots with eleven or fewer crossings. At the time I write, the classification has been extended through 14 crossings.

There are many strange ways to label the crossings of a knot, then derive an algebraic expression that is an invariant for all possible diagrams of that knot. One of the earliest of such techniques produces what is called a knot's Alexander polynomial, named after the American mathematician James W. Alexander who discovered it in 1928. Conway later found a beautiful new way to compute a "Conway polynomial" that is equivalent to the Alexander one.

For the unknotted knot with no crossings the Alexander polynomial is 1. The expression for the trefoil knot of three crossings is $x^2 - x + 1$, regardless of its handedness. The figure-eight knot of four crossings has the polynomial $x^2 - 3x + 1$. The square knot, a product of two trefoils, has an Alexander polynomial of $(x^2 - x + 1)^2$, the square of the trefoil's expression. Unfortunately, a granny knot has the same polynomial. If two knot diagrams give different polynomials, they are sure to be different knots, but the converse is not true. Two knots may have the same polynomial yet not be the same. Finding a way to give any knot an expression that applies to all diagrams of that knot, and only that knot, is the major unsolved problem in knot theory.

Although there are tests for deciding whether any given knot is trivial, the methods are complex and tedious. For this reason many problems that are easy to state are not easy to resolve except by working empirically with rope models. For instance, is it possible to twist an elastic band around a cube so that each face of the cube has an under-over crossing as shown in Figure 28? To put it another way, can you tie a cord around a cube in this manner so that if you slip the cord off the cube, the cord will be unknotted?

Note that on each face the crossing must take one of the four forms depicted in the illustration. This makes $4^6 = 4,096$ ways to wrap the cord. The wrapping can be diagrammed as a 12-knot, with six pairs of crossings, each pair of which can have one of four patterns. The problem was first posed by Horace W. Hinkle in *Journal of Recreational Mathematics* in 1978. In a later issue (Vol. 12, No. 1, pages 60-62; 1979-80) Karl Scherer showed how symmetry considerations reduce the number of essentially different wrappings to 128. Scherer tested

Figure 28

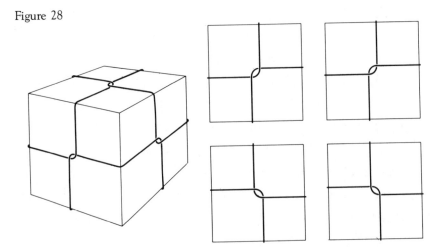

each wrapping empirically and found that in every case the cord is knotted. This has yet to be confirmed by others, and no one has so far found a simpler way to attack the problem. The impossibility of getting the desired wrapping with an unknotted cord seems odd, because it is easy to twist a rubber band around a cube to put the under–over crossings on just two or four faces (all other faces being straight crossings), and seemingly impossible to do it on just one face, three faces, or five faces. One would therefore expect six to be possible, but apparently it is not. It may also be impossible to get the pattern even if two, three, or four rubber bands are used.

Figure 29 depicts a delightful knot-and-link puzzle that was sent to me recently by its inventor, Majunath M. Hegde, a mathematics student

Figure 29

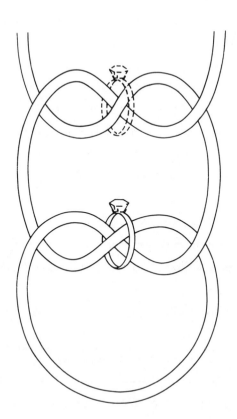

in India. The rope's ends are tied to a piece of furniture, say a chair. Note that the two trefoil knots form a granny. The task is to manipulate the rope and ring so that the ring is moved to the upper knot as is indicated by the broken line. All else must remain identical.

It is easy to do if you have the right insight. Of course, the rope must not be untied from the chair, nor are you allowed to open a knot and pass the chair through it. It will help if you think of the ends of the rope as being permanently fastened to a wall.

The trick of dissolving or creating knots by passing a person through a loop was actually used by fake mediums in the days when it was fashionable to relate psychic phenomena to the fourth dimension. Knots in closed curves are possible only in 3-space. In 4-space all knots dissolve. If you could toss an unknotted loop of rope to a creature in 4-space, it could tie any knot in the loop and toss it back to you with the knot permanently formed. There was a popular theory among physicists who believed in spiritualism that mediums had the power to move objects in and out of higher spaces. Some mediums, such as the American mountebank Henry Slade, exploited this theory by pretending to put knots into closed loops of cord. Johann C. F. Zöllner, an Austrian physicist, devoted an entire book to Slade and hyperspace. Its English translation, *Transcendental Physics* (Arno Press, 1976), is worth reading as striking testimony to the ease with which an intelligent physicist can be gulled by a clever conjurer.

Scientists are still being taken in by tricks involving knots and links. Psychic investigators William Cox and John Richards have recently been exhibiting a stop-action film that purports to show two leather rings becoming linked and unlinked inside a fish tank. "Later examination showed no evidence that the rings were severed in any way," wrote *National Enquirer* when it reported this miracle on October 27, 1981. I was reminded of an old conjuring stage joke. The performer announces that he has magically transported a rabbit from one opaque box to another. Then before opening either box he says that he will magically transport the rabbit back again.

It is easy, by the way, to fabricate two linked "rubber bands." Just draw them linked on the surface of a baby's hollow rubber teething ring and carefully cut them out. Two linked wood rings, each of a different wood, can be carved if you insert one ring into a notch cut into a tree, then wait many years until the tree grows around and through it. Because the trefoil is a torus knot, it too is easily cut from a teething ring.

The trick I am about to describe was too crude for Slade, but less clever mediums occasionally resorted to it. You will find it explained, along with other knot-tying swindles, in Chapter 2 of Hereward Carrington's *The Physical Phenomena of Spiritualism, Fraudulent and Genuine* (H. B. Turner & Co., Boston, 1907). One end of a very long piece of rope is tied to the wrist of one guest and the other end is tied to the wrist of another guest. After the seance, when the lights are turned on, several knots are in the rope. How do they get there?

The two guests stand side by side when the lights go out. In the dark the medium (or an accomplice) makes a few large coils of rope, then passes them carefully over the head and body of one of the guests. The coils lie flat on the floor until later, when the medium casually asks that guest to step a few feet to one side. This frees the coils from the person, allowing the medium to pull them into a sequence of tight knots at the center of the rope. Stepping to one side seems so irrelevant to the phenomenon that no one remembers it. Ask the guest himself a few weeks later whether he changed his position, and he will vigorously and honestly deny it.

Roger Penrose, the British mathematician and physicist, once showed me an unusual trick involving the mysterious appearance of a knot. Penrose invented it when he was in grade school. It is based on what in crocheting, sewing, and embroidery is called a chain stitch. Begin the chain by trying a trefoil knot at one end of a long piece of heavy cord or thin rope and hold it with your left hand as in step 1 in Figure 30. With your right thumb and finger take the cord at A and pull down a loop as in step 2. Reach through the loop, take the cord at B, and pull down another loop (step 3). Again reach forward through the lowest loop, take the cord at D, and pull down another loop (step 4). Continue in this way until you have formed as long a chain as possible.

With your right hand holding the lower end of the chain, pull the chain taut. Ask someone to select any link he likes and then pinch the link between his thumb and forefinger. Pull on both ends of the cord. All links dissolve, as expected, but when he separates his finger and thumb, there is a tight knot at precisely the spot he pinched!

A few years ago Joel Langer, a mathematician at Case Western Reserve University, made a remarkable discovery. He found a way of constructing what he calls "jump knots" out of stainless-steel wire. The wire is knotted and then its ends are bonded. When it is manipulated properly, it can be pressed flat to form a braided ring. Release pressure

Figure 30

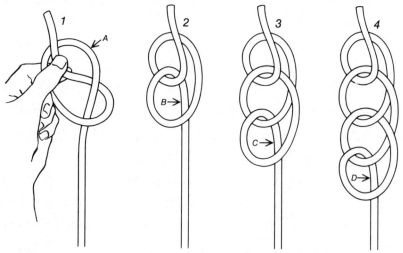

on the ring; tension in the wire causes it to spring suddenly into a symmetrical three-dimensional shape. It is now a frustrating puzzle to collapse the wire back to its ring form.

In 1981 Langer and his associate Sharon O'Neil formed a company they called Why Knots. From it you can obtain three handsome jump knots: the Figure Eight, the Chinese Button Knot, and the Mathematician's Loop. When you slide one of these wire knots out of its square envelope, it pops into an elegant hanging ornament. The figure eight is the easiest to put back into its envelope. The Chinese button knot (so called because it is a form widely used in China for buttons on nightclothes) is more difficult. The mathematician's loop is the most difficult.

Langer tells me that anyone in the U.S. can get his three jump knots by sending $10.50 to Why Knots, P.O. Box 635, Aptos, CA 95003. These shapes make it easier to understand how the 18th-century physicists could have developed a theory, respectable in its day, that molecules are different kinds of knots into which vortex rings of ether (today read "space-time") get themselves tied. Indeed, it was just such speculation that led the Scottish physicist Peter Guthrie Tait to study topology and conduct the world's first systematic investigation of knot theory.

Answers

Figure 31 shows how a square knot can be changed to an alternating knot of six crossings. Simply flip dotted arc *a* over to make arc *b*.

Figure 32 shows one way to solve the ring-and-granny-knot puzzle. First make the lower knot small, then slide it (carrying the ring with it) up and through the higher knot (*a*). Open it. Two trefoil knots are now side by side (*b*). Make the ringless knot small, then slide it through and down the other knot. Open it up and you have finished (*c*).

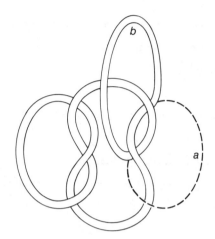

Figure 31

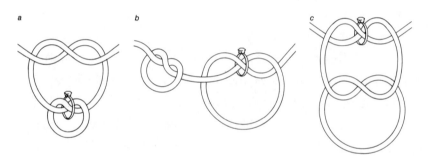

Figure 32

ADDENDUM

Enormous advances in knot theory have been made since this chapter was written in 1983, and knot theory is now one of the most exciting and active branches of mathematics. Dozens of new polynomials for classifying knots have been discovered. One is called the Homfly after the last initials of its six independent discoverers. The most significant new expression is the Jones polynomial found in 1984 by the New Zealand mathematician Vaughan F. R. Jones, now at the University of California, Berkeley. It has since been improved and generalized by Louis Kauffman and others. Although these new polynomials are surprisingly simple and powerful, no one has yet come up with an algebraic technique for distinguishing all knots. Knots with different polynomials are different, but it is still possible that two distinct knots will have the same expression.

The Alexander polynomial does not decide between mirror-image knots, and as we have seen, it does not distinguish the square knot from the granny. The Jones polynomial provides both distinctions. So far, it is not clear just why the Jones and the other new polynomials work. "They're magic" is how Joan Birman, a knot expert at Barnard College, put it.

The most amazing development in recent knot theory was the discovery that the best way to understand the Jones polynomial was in terms of statistical mechanics and quantum theory! Sir Michael Atiyah of Cambridge University, was the first to see these connections, then Edward Witten, at the Institute for Advanced Study in Princeton, did the pioneer work in developing the connections. Knot theory now has surprising applications to superstrings, a theory that explains basic particles by treating them as tiny loops, and to quantum field theory. There is intense interaction between today's physicists and topologists. Discoveries in physics are leading to new discoveries in topology, and vice versa. No one can predict where it will all lead.

Another unexpected application of knot theory is in broadening our understanding of the structure and properties of large molecules such as polymers, and especially the behavior of DNA molecules. DNA strands can become horribly knotted and linked, unable to replicate until they are untied or unlinked by enzymes. To straighten out a DNA strand, enzymes have to slice them so they can pass through themselves or another strand, then splice the ends together again. The number of times this must occur to undo a knot or linkage of DNA determines the speed with which the DNA unknots or unlinks.

There is a delightful three-color test for deciding if a knot diagram represents a knot. Draw the diagram, then see if you can color its "arcs" (portions of the line between two crossings) with three colors so that either all three colors meet at each crossing, or there is only one color at

each crossing, and provided at least one crossing shows all three colors. If
you can do this, the line is knotted. If you can't, the line may or may not
be knotted. The three-coloring can also be used to prove that two knots
are different.

In 1908 the German mathematician Heinrich Tietze conjectured that
two knots are identical if and only if their complements—the topological
structure of the space in which they are embedded—are identical. His conjecture was proved in 1988 by two American mathematicians, Cameron M.
Gordon and John E. Luecke. A knot's complement is a structure in 3-space,
in contrast to the knot which is one-dimensional. Its topological structure is
more complicated than the knot's, but of course it contains complete information about the knot. The theorem fails for links. Two links that are not
the same can have identical complements.

Associated with each knot's complement is a group. Like the polynomials, which can be extracted from the group, two knots can have
the same group yet not be the same knots. An anonymous poet
summed up the situation this way in the British periodical *Manifold*
(Summer 1972):

A knot and

another

knot may

not be the

same knot, though

the knot group of

the knot and the

other knot's

knot group

differ not; BUT

if the knot group

of a knot

is the knot group

> of the not
> knotted
> knot,
> the knot is
> not
> knotted.

The American philosopher Charles Pierce, in a section on knots in his *New Elements of Mathematics* (Volume 2, Chapter 4), shows how the Borromean rings (three rings linked in such a way that although they can't be separated, no two rings are linked) can be cut from a three-hole torus. Peirce also shows how to cut the figure-eight knot and the bowline knot from a two-hole torus.

Richard Parris called attention to the fact that not all of the 4,096 ways to wrap string around the cube, in the problem I posed, are knots. Most of them are links of two, three, or four separate loops.

References

Books

INTRODUCTION TO KNOT THEORY. R. H. Crowell and R. H. Fox. Blaisdell, 1963; Springer-Verlag, 1977.

KNOTS AND LINKS. Dale Rolfsen. Publish or Perish, 1976, Second edition, 1990.

ON KNOTS. Louis Kauffman. Princeton University Press, 1987.

NEW DEVELOPMENTS IN THE THEORY OF KNOTS. Toshitake Kohno. World Scientific, 1990.

THE GEOMETRY AND PHYSICS OF KNOTS. Michael Atiyah. Cambridge University Press, 1990.

KNOTS AND PHYSICS. Louis Kauffman. World Scientific, 1991.

KNOT THEORY. Charles Livingston. Mathematical Association of America, 1993.

THE KNOT BOOK. Colin C. Adams. Freeman, 1994.

THE HISTORY AND SCIENCE OF KNOTS. J. C. Turner and P. van de Griend. World Scientific, 1996.

Papers

Out of hundreds of papers on knot theory published since 1980, I have selected only a few that have appeared since 1990.

UNTANGLING DNA. De Witt Summers in *The Mathematical Intelligencer*, Vol. 12, pages 71–80; 1990.

KNOT THEORY AND STATISTICAL MECHANICS. Vaughan F. R. Jones, in *Scientific American*, pages 98–103; November 1990.

RECENT DEVELOPMENTS IN BRAID AND LINK THEORY. Joan S. Birman in *The Mathematical Intelligencer*, Vol. 13, pages 57–60; 1991.

KNOTTY PROBLEMS—AND REAL-WORLD SOLUTIONS. Barry Cipra in *Science*, Vol. 255, pages 403–404; January 24, 1992.

KNOTTY VIEWS. Ivars Peterson in *Science News*, Vol. 141, pages 186–187; March 21, 1992.

KNOTS, LINKS AND VIDEOTAPE. Ian Stewart in *Scientific American*, pages 152–154; January 1994.

BRAIDS AND KNOTS. Alexey Sosinsky in *Quantum*, pages 11–15; January/February 1995.

HOW HARD IS IT TO UNTIE A KNOT? William Menasco and Lee Rudolph in *American Scientist*, Vol. 83, pages 38–50; January/February 1995.

THE COLOR INVARIANT FOR KNOTS AND LINKS. Peter Andersson in *American Mathematical Monthly*, Vol. 102, pages 442–448; May 1995.

GEOMETRY AND PHYSICS. Michael Atiyah in *The Mathematical Gazette*, pages 78–82; March 1996.

KNOTS LANDING. Robert Matthews in *New Scientist*, pages 42–43; February 1, 1997.

6

M-Pire Maps

"I know by the color. We're right over Illinois yet. [Huckleberry Finn is speaking to Tom; they are on a balloon trip.] And you can see for yourself that Indiana ain't in sight. . . .

What's the color got to do with it?

It's got everything to do with it. Illinois is green, Indiana is pink. . . . I've seen it on the map, and it's pink."

—Mark Twain, *Tom Sawyer Abroad*

In 1976 Wolfgang Haken and Kenneth Appel of the University of Illinois at Urbana-Champaign announced they had finally laid to rest the famous four-color-map problem. As the reader surely knows, this renowned conjecture in topology asserts that four colors are both sufficient and necessary for coloring all maps drawn on a plane or sphere so that no two regions that "touch" (share a segment of a boundary) are the same color. Haken and Appel, with the assistance of John Koch, proved that the conjecture is true by a method that made unpre-

cedented use of computers. Their proof is an extraordinary achievement, and when their account of it was published in 1977, the Urbana
post office proudly added to its postmark "Four colors suffice." To most
mathematicians, however, the proof of the four-color conjecture is deeply
unsatisfying.

For more than a century topologists either suspected that a
counterexample to the four-color conjecture (that is, a complex map
requiring five colors) could be devised or trusted that a simple, elegant
proof of the conjecture could be found. Although the conjecture is now
known to be true, its proof is buried in printouts that resulted from
1,200 hours of computer time. The task of verifying the accuracy of
these results is so horrendous that only a small number of experts have
had the time, fortitude, and skill to even attempt it. So far, however, all
who have done so have attested to the proof's validity.

In an article titled "The Four-Color Problem and Its Philosophical
Significance," published in *The Journal of Philosophy* (Vol. 76, No. 2,
pages 57–83; February 1979), Thomas Tymoczko argues that this kind
of lengthy computer proof injects an empirical element into mathematics. No mathematician, he writes, has seen a proof of the four-color
theorem, nor has anyone seen a proof that the work of Haken and
Appel is, in fact, a proof. What mathematicians have seen instead is a
program for attacking the problem by computer along with the results
of an "experiment" performed on a computer. Tymoczko believes such
a "proof" blurs the distinction between mathematics and natural science and lends credibility to the opinions of those contemporary philosophers of science, such as Hilary Putnam who see mathematics as a
"quasi-empirical" activity.

There is, of course, something to this viewpoint. All mathematical
proofs are the work of human beings, and when proofs are extremely
complex, human error is always a possibility. The validity of a difficult
proof rests on a consensus among experts, who may, after all, be mis-

taken. There is a striking instance of this in the early history of the four-color theorem. Alfred Bray Kempe, an English mathematician, published what he said was a proof of the theorem in 1879, and for about a decade mathematicians assumed that the problem had been solved. Then in 1890 Percy John Heawood, another English mathematician, pointed out a fatal flaw in Kempe's reasoning.

My purpose here is not to argue the question of whether there is a sharp line separating "analytic" truth from "synthetic" truth. I shall say only that I think Tymoczko greatly overestimates the relevance of modern computers to this old controversy. All calculations are empirical in the trivial sense that they involve the carrying out of an experiment with symbols, either in the head, with pencil and paper, or with the aid of a machine. The fact that with electronic computers, which are now essential for difficult calculations, mistakes can be made by both hardware and software differs in no essential way from the fact that mistakes can be made by a person multiplying two large numbers on an abacus. It seems to me a misuse of language to say that the possibility of such errors makes the truth of the multiplication table empirical and therefore a mistake to take this kind of inescapable error as an example of the fallibility of natural science.

Still, the Haken–Appel proof of the four-color theorem is certainly unsatisfying in that no one can call it simple, beautiful, or elegant. Haken and Appel both think it unlikely that a proof can be found that does not require an equally intensive application of computers, but of course there is no way to be sure. If there is no simpler proof, the Haken–Appel proof is indeed something new in the degree to which it relies on computer technology.

This situation is ably discussed by Benjamin L. Schwartz in a book he edited titled *Mathematical Games and Solitaires*. Published in 1979 by the Baywood Publishing Company of Farmingdale, N.Y., the book (which I enthusiastically recommend) is a choice selection of articles

from *Journal of Recreational Mathematics.* In Schwartz's introduction to
a section on the four-color problem he writes: "So one may ask, have
Haken and Appel really proved what they claim?. . . Personally I believe
they have. . . . But the trial period is still not over. Others will have to
check every step. And since [most] of the steps were carried out in
hundreds of hours of high-speed computer operation, that checking
will be a big job. At this writing no one had done it. New computer
code will have to be written, perhaps for another computer. . . . Will a
whole set of other stubborn mathematical problems. . . begin to yield to
the new method of massive computational support? Or is this a fluke
case that will have no lasting impact? This proof of the four-color theorem introduces a new era in mathematics, and no one knows where it
will lead."

In December of 1976 G. Spencer-Brown, the maverick British mathematician, startled his colleagues by announcing he had a proof of the
four-color theorem that did not require computer checking. Spencer-
Brown's supreme confidence and his reputation as a mathematician
brought him an invitation to give a seminar on his proof at Stanford
University. At the end of three months all the experts who attended the
seminar agreed that the proof's logic was laced with holes, but Spencer-

Brown returned to England still sure of its validity. The "proof" has not yet been published.

Spencer-Brown is the author of a curious little book called *Laws of Form*, which is essentially a reconstruction of the propositional calculus by means of an eccentric notation. The book, which the British mathematician John Horton Conway once described as beautifully written but "content-free," has a large circle of counterculture devotees. Incidentally, after Brown's announcement that he had proved the four-color theorem was reported in newspapers around the world the *Vancouver Sun* for January 17, 1977, printed a letter from a woman in British Columbia. Brown could not have proved the theorem, she wrote, because in April 1975, *Scientific American* had printed a map that required five colors. She was referring to a map that appeared in my column as an April Fools' joke!

While topologists go on with their search for a simple proof of the four-color theorem, some are also working on two fascinating but little-known generalizations of the problem that are still unsolved. In what follows I shall draw heavily on a private communication from Herbert Taylor. Formerly a mathematician at California State University at Northridge and at the Jet Propulsion Laboratory of the California Institute of Technology, he is currently studying electrical engineering with Solomon W. Golomb at the University of Southern California. He was also once rated one of the world's top three non-Oriental go players.

As Taylor points out, one way to generalize the four-color problem is to consider a map on which each country, or area to be colored, consists of m disconnected regions. If all regions of a single country must be the same color, what is the smallest number of colors necessary for coloring any such map so that no two regions of like color share a common border? Taylor calls this question the m-pire problem and the number of colors required the m-pire chromatic number.

If m equals 1 (that is, if each country consists of only one region), the problem is equivalent to the four-color problem, and Haken and Appel established that the chromatic number is 4. If m equals 2 (think of each country as having one colony with the same color as the country), the chromatic number is 12. Surprisingly, this result was presented by Heawood in 1890 in the same paper in which he demolished Kempe's proof of the four-color theorem. In other words, the solution to the m-pire problem for the case $m = 2$ was obtained long before the solution for the case $m = 1$. In Heawood's proof he first showed that for any positive integer m the required number of colors for an m-pire map is no more than $6m$. Then he exhibited a "2-pire" map that required 6×2, or 12, colors, a map he said he "obtained with much difficulty in a more or less empirical [m-pirical?] manner." That map is shown in Figure 33 (see color plate).

Note that Heawood's map has no symmetry. Taylor found a fairly symmetrical version (which can be obtained from the map shown at the top of Figure 34 by shrinking the lettered regions to points), but the most symmetrical map was devised recently by Scott Kim, a graduate student at Stanford University. Kim's beautiful map is shown in Figure 35 (see color plate). As Heawood remarked of his own map: "What essential variety there might be in such an arrangement of 12 two-division countries. . . is a curious problem, to which the one figure obtained does not afford much clue."

Heawood was convinced that $6m$ provides the chromatic number for all m-pire maps. Examine Heawood's map or Kim's for the case $m = 2$, and you will see that each 2-pire touches all the others, thereby proving that 12 colors are necessary. Heawood believed that for every m there exists a similar pattern of $6m$ regions, with each m-pire touching all the others. Taylor recently proved that this conjecture is true when m equals 3, using the map requiring 6×3, or 18, colors (Figure 36, see color plate). Note that only two regions on the map are numbered 18

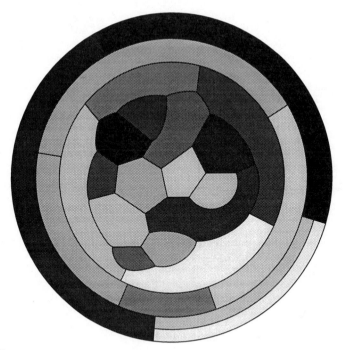

Figure 33

and colored yellow. The third region of this 3-pire is disconnected from the rest of the map and can be anywhere on the plane.

Taylor later confirmed Heawood's conjecture for the case $m = 4$ by constructing the two-part map requiring 6×4, or 24, colors (Figure 34). Think of the two parts of the map as being two hemispheres of the same sphere. (Any map on a spherical surface can be converted into a topologically equivalent planar map by puncturing the surface inside any region and then stretching the hole until the map can be laid flat.) Note that each 4-pire on the map touches all the others, proving that 24 colors are necessary in the 4-pire problem. Both of these results are published here for the first time. Heawood's conjecture remains unverified for the case $m = 5$ and all higher values of m. For maps drawn on the surface of a torus, however, Taylor solved the m-pire problem. He has submitted a note to

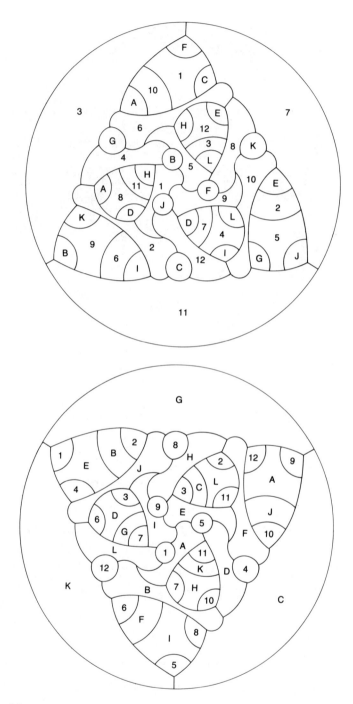

Figure 34

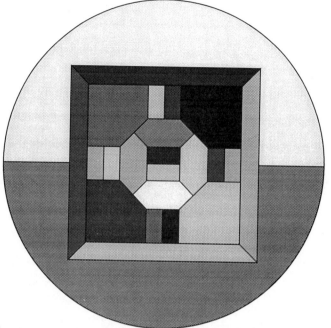

Figure 35

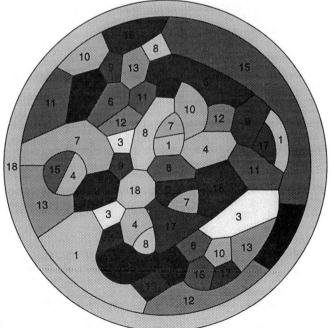

Figure 36

Journal of Graph Theory titled "The *m*-pire Chromatic Number of the Torus Is $6m + 1$." The problem remains open for all toruses with more than one hole.

In 1959, in a German book dealing with graph-coloring problems, Gerhard Ringel posed another problem closely related to the *m*-pire problem. Assume that Mars has been colonized by the nations of the earth and that each nation has one home territory on this planet and one colony on Mars. Each region is simply connected (without holes), and each colony is the same color as its parent nation. Once again the problem is to find the minimum number of colors that will color all possible maps on the two spheres so that no two regions of the same color touch at more than single points. Since maps on spheres are equivalent to planar maps, the same problem can be formulated in terms of two separate maps on the plane.

Ringel showed that the chromatic number for all two-sphere maps is either 8, 9, 10, 11, or 12. The upper bound of 12 is derived from Heawood's upper bound for the *m*-pire problem of $6m$ as follows. Suppose a pair of maps requires more than 12 colors. It would then be possible to convert them into planar maps and join them to create a 2-pire map requiring more than 12 colors, thereby violating the proved upper bound of $6m$.

Ringel guessed the chromatic number for the Earth–Mars maps to be 8, a hypothesis that was strengthened in 1962 when Joseph Battle, Frank Harary, and Yukihiro Kodama showed that a two-sphere map could not be constructed with nine 2-pires so that each 2-pire touched all the others. In 1974, however, Thom Sulanke, then a student at Indiana University, sent Ringel the pair of maps shown in Figure 37. These maps too are published here for the first time. If you try to color the eleven 2-pires so that both of the regions having the same number are given the same color, you will find that nine colors are needed! Thus to color the Earth–Mars maps nine colors are necessary and 12

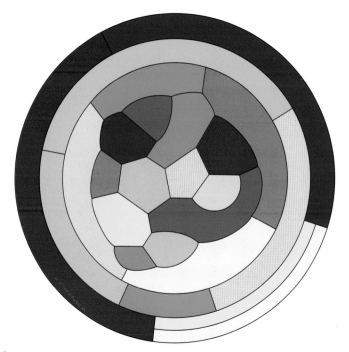

Figure 33

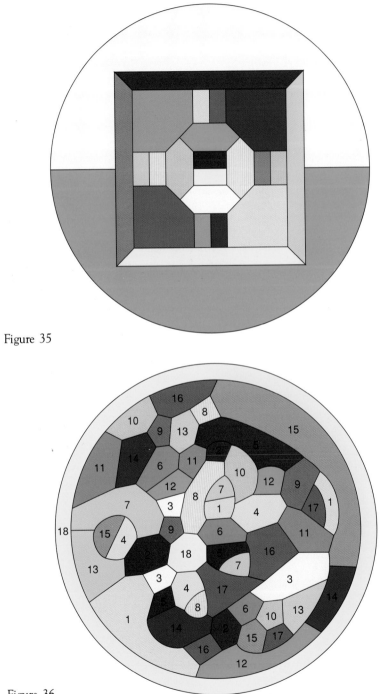

Figure 35

Figure 36

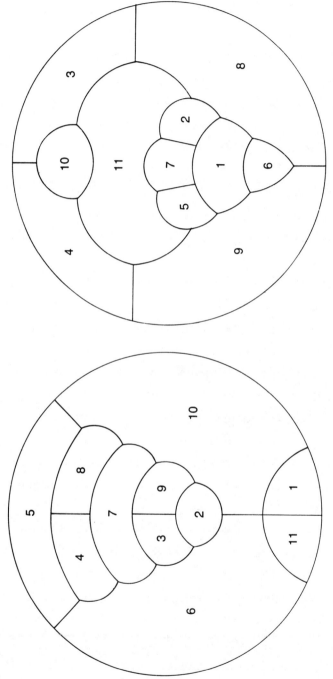

Figure 37

are sufficient. No one yet knows if such a pair of maps can be constructed that require 10, 11, or 12 colors.

It is also possible to combine the 2-sphere problem with the m-pire problem. For example, suppose m equals 4 and each sphere is a map on which each country has just two regions. If you think of the patterns in Figure 37 as being two separate maps, one on the earth and the other on Mars, they prove that for the case $m = 4$ the number of colors required is 24. We know that 24 are enough as well because Heawood's upper bound of $6m$ also applies here. Hence the problem is solved. Taylor conjectures that for every positive even-integer m there is a map of $6m$ m-pires on a surface consisting of $m/2$ spheres such that each m-pire has two of its m parts on each sphere and each m-pire touches all the others.

I conclude with a delightful coloring puzzle involving the U.S. Ignoring Hawaii and the disconnected parts of states such as the islands that belong to New York and California, note that nowhere on the map of the 49 states of the continental U.S. is there a place where four states all mutually share borders. (The same is not true of other countries. For example, Switzerland has four cantons that are mutual neighbors: Solothurn lies at the center of the configuration, and it is surrounded by Aargau, Basel, and Bern.) This situation suggests an intriguing question: Is it possible to color the 49 states with three colors instead of four?

Another way to view this possibility is to consider the Four Color Puzzle Game, marketed in 1979 by Knots, Inc. (2425 Third Street, San Francisco, CA 94107). People who buy the game (for $6.95 postpaid) are given two jigsaw puzzle maps of the continental U.S. In each puzzle each state is represented by a single piece, and the two pieces in the game representing each state are different colors. The task is to choose pieces to make a four-color map of the U.S. in which no neighboring states are the same color. (As in the four-color theorem, states of the same color may touch at a single point.) To restate our question: Is it possible that Knots, Inc., could have used only three colors for their puzzle pieces and asked for a three-color map of the U.S.?

The answer is no, but most people find it annoyingly difficult to prove. Can the reader give a simple proof that the U.S. map requires four colors?

Answers

The map-coloring problem is from Howard P. Dinesman's collection of brainteasers *Superior Mathematical Puzzles, with Detailed Solutions* (Simon and Schuster, 1968). It can be answered as follows. Nevada is surrounded by a ring of five states: Oregon, Idaho, Utah, Arizona, and California. Color Nevada with color 1. If only three colors are used, each state in the ring must be colored with either color 2 or color 3 to avoid conflict with Nevada, and these two colors must alternate around the ring. Since the ring consists of an odd number of states (five), however, there is no way to avoid giving two adjacent states the same color. Therefore a fourth color is necessary. The same applies to the ring of five states that surround West Virginia, and the ring of seven around Kentucky.

This property of rings that consist of an odd number of regions plays a basic role in map-coloring theory. Consider how it applies to L. Frank Baum's land of Oz. Oz is made up of five regions, each region with a dominant landscape color: the green Emerald City is surrounded by a ring consisting of the yellow Winkie country, the red Quadling country, the blue Munchkin country and the purple Gillikin country. Surrounding all of Oz is the great Deadly Desert. Because four is an even number, a map of Oz can be colored with three colors, but of course no Oz cartographer would use fewer than five for Oz and a sixth for the surrounding desert.

ADDENDUM

English Professor James Kirkup, at Kyoto University, in Japan, sent a letter about my *m*-pire map coloring column to the *Houseman Society Journal*. Published in Volume 7 (1981), pages 83–84, the letter begins:

> Dear Sir,
>
> I always enjoy the Mathematical Games published in *The Scientific American* and I was particularly interested in the issue of February, 1980, which discussed the colouring of unusual maps. The epigraph from Mark Twain's *Tom Sawyer Abroad* brought to mind the celebrated line by A. E. Houseman from "Bredon Hill":
>
> > Here of a Sunday morning
> > My love and I would lie,
> > *And see the coloured counties,*
> > And hear the larks so high
> > About us in the sky.

Kirkup italicized the third line. He goes on to wonder if Houseman was familiar with the four-color map theorem.

Ian Stewart, in his *Scientific American* article listed as a reference, reports that for 3-pire maps—say one on the Earth, another on the moon, and a third on Mars—the optimal number of colors is 16, 17, or 19. For $m = 4$ or higher, the number is $6m$, $6m - 1$, or $6m - 2$.

Scott Kim sent me another remarkable version of his 2-pire map, here reproduced as Figure 38. It will fold into a cube with partially truncated corners. "The hexagons are the truncations," Kim writes, "and the squares are concentric with the cube's faces. The symmetry is pretty (it makes a great model), but a little misleading. It is true that each empire consists of a rectangle and a hexagon, but there are two distinct ways they are paired. Empires 1, 2, 4, 7, 8, 9 are one type. The only symmetry that preserves empire type is a rotation around the axis that connects the intersection point of hexagons 1, 4, and 8, and the intersection point of hexagons 6, 10, and 11."

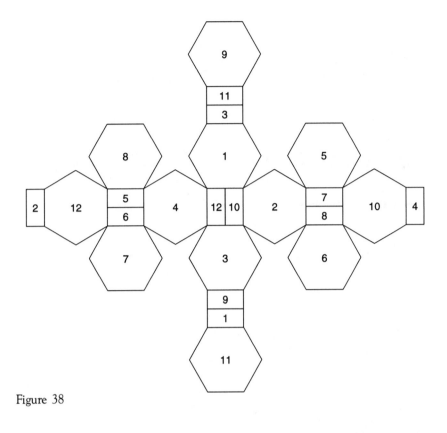

Figure 38

References

THE FOUR-COLOR PROBLEM. Oystein Ore. Academic Press, 1967.

THE FOUR-COLOR PROBLEM AND THE FIVE-COLOR THEOREM. Anatole Beck, Michael Bleicher, and Donald Crowe in *Excursions Into Mathematics*, Section 6. Worth, 1969. Heawood's original 2-pire map is reproduced in color as a frontispiece, and there is a photograph of Heawood on page 55.

THE SOLUTION OF THE FOUR-COLOR-MAP PROBLEM. Kenneth Appel and Wolfgang Haken in *Scientific American*. Vol. 237, pages 108–121; October 1977.

HEAWOOD'S EMPIRE PROBLEM. R. Jackson and G. Ringel in *Journal of Combinatorial Theory*, Series B, pages 168–178; 1985.

PEARLS IN GRAPH THEORY: A COMPREHENSIVE INTRODUCTION. Nora Hatsfield and Gerhard Ringel. Academic Press, 1990.

THE RISE AND FALL OF THE LUNAR M-PIRE. Ian Stewart in *Scientific American*, pages 120–121; April 1993.

COLORING ORDINARY MAPS, MAPS OF EMPIRES, AND MAPS OF THE MOON. Joan P. Hutchinson in *Mathematics Magazine*, Vol. 66, pages 211–226; October 1993. Twenty-nine references are listed at the end of this excellent paper.

7

Directed Graphs and Cannibals

Stranger in car: "How do I get
to the corner of Graham Street and
Harary Avenue?"

Native on sidewalk: "You can't
get there from here."

n graph theory a graph is defined as any set of points joined by lines,
and a simple graph is defined as one that has no loops (lines that join
a point to itself) and no parallel lines (two or more lines joining the
same pair of points). If an arrowhead is added to each line of a graph,
giving each line a direction that orders its end points, the graph becomes a directed graph, or digraph for short. Directed lines are called
arcs. Digraphs are the subject here, and the old joke quoted above is

appropriate because on some digraphs it is actually impossible to get from one specified point to another.

A digraph is called complete if every pair of points is joined by an arc. For example, a complete digraph for four points is shown in Figure 39 (left). The figure at the right is the adjacency matrix of the digraph, which is constructed as follows. Think of the digraph as a map of one-way streets. Starting at point A, it is possible to go directly only to point B, a fact that is indicated in the top row of the matrix (the row corresponding to A) by putting a 1 in the column corresponding to B and a 0 in all the other columns. The remaining rows of the adjacency matrix are determined in the same way, so that the matrix is combinatorially equivalent to the digraph. It follows that given the adjacency matrix it is easy to construct the digraph.

Other important properties of digraphs can be exhibited in other kinds of matrixes. For example, in a distance matrix each cell gives the smallest number of lines that form what is called a directed path from one point to another, that is, a path that conforms to the arrowheads on the graph and does not visit any point more than once. Similarly, the cells of a detour matrix give the number of lines in the longest directed path between each pair of points. And a reachability matrix

Figure 39

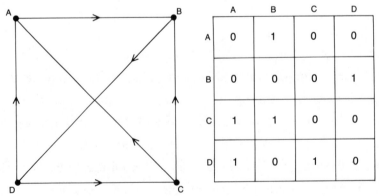

indicates (with 0s and 1s) whether a given point can be reached from another point by a directed path of any length. If every point is reachable from every other point, the digraph is said to be strongly connected. Otherwise there will be one or more pairs of points for which "you can't get there from here."

The following theorem is one of the most fundamental and surprising results about complete digraphs: No matter how the arrowheads are placed on a complete digraph, there will always be a directed path that visits each point just once. Such a path is called a Hamiltonian path after the Irish mathematician William Rowan Hamilton. Hamilton marketed a puzzle game based on a graph equivalent to the skeleton of a dodecahedron in which one task was to find all the paths that visit each point just once and return to the starting point. A cyclic path of this type is called a Hamiltonian circuit. (Hamilton's game is discussed in Chapter 6 of my *Scientific American Book of Mathematical Puzzles & Diversions*.)

The complete-digraph theorem does not guarantee that there will be a Hamiltonian circuit on every complete digraph, but it does ensure that there will be at least one Hamiltonian path. More surprisingly, it turns out that there is always an odd number of such paths. For example, on the complete digraph in Figure 40 there are five Hamiltonian paths: *ABDC*, *BDCA*, *CABD*, *CBDA*, and *DCAB*. All but one of them (*CBDA*) can be extended to a Hamiltonian circuit.

The theorem can be expressed in other ways, depending on the interpretation given the graphs. For example, complete digraphs are often called tournament graphs because they model the results of the kind of round-robin tournaments in which each player plays every other player once. If *A* beats *B*, a line goes from *A* to *B*. The theorem guarantees that whatever the outcome of a tournament is all players can be ranked in a column so that each player has defeated the player immediately below him. (It is assumed here that, as in tennis, no game can end

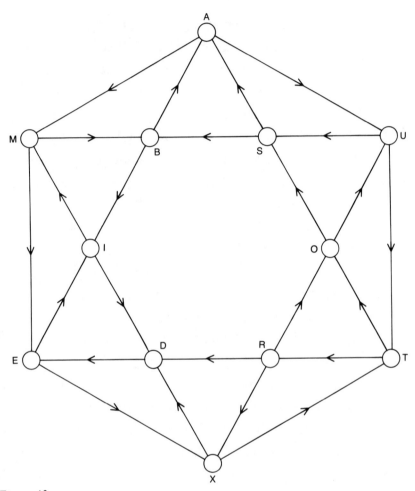

Figure 40

in a draw. If a game did allow draws, they would be represented by
undirected lines and the graph would be called a mixed graph. Mixed
graphs can always be converted into digraphs by replacing each undirected line with a pair of directed parallel lines going in opposite directions.)

Tournament graphs can be applied to represent many situations
other than tournaments. Biologists have used the graphs to diagram

the pecking order of a flock of chickens or, more generally, to diagram the structure that any other kind of pairwise dominance relation imposes on a population of animals. Social scientists have used
the graphs for modeling dominance relations among people or groups
of people. Tournament graphs provide a convenient means of modeling a person's pairwise preferences for any set of choices, such as
brands of coffee or candidates in an election. In all these cases the
theorem guarantees that the animals, people, or objects in question
can always be ordered in a linear chain by means of the one-way
relation.

The theorem is tricky to prove, but to convince yourself of its validity try labeling a complete graph of n points so that no Hamiltonian
path is created. The impossibility of the task suggested the following
pencil-and-paper game to the mathematician John Horton Conway. Two
players take turns adding an arrowhead to any undirected line of a
complete graph, and the first player to complete a Hamiltonian path
loses. The theorem ensures that the game cannot be a draw. Conway
finds the play is not interesting unless there are seven or more points in
the graph.

The digraph in Figure 40 appeared as a puzzle in the October 1961
issue of the Cambridge mathematical annual *Eureka*. Although it is not
a complete digraph, it has been cleverly labeled with arrowheads so that
it has only one Hamiltonian circuit. Think of the graph as a map of
one-way streets. You want to start at A and drive along the network,
visiting each intersection just once before returning to A. How can it be
done? (Hint: The circuit can be traced by a pencil held in either hand.)

Digraphs can provide puzzles or be applied as tools for solving
puzzles in innumerable ways. For example, the graphs serve to model
the ways a flexagon flexes, and they are valuable in solving moving-
counter and sliding-block puzzles and chess-tour problems. Probability
questions involving Markov chains often yield readily to a digraph analysis,

and winning strategies for two-person games in which each move alters the state of the game are frequently found by exploring a digraph of all possible plays. In principle even the game of chess could be "solved" by examining its digraph, but the graph would be so enormous and so complex that it will probably never be drawn.

Digraphs are extremely valuable in the field of operations research, where they can be applied to solve complicated scheduling problems. Consider a manufacturing process in which a certain set of operations must be performed. If each operation requires a fixed amount of time to perform and certain operations must be completed before others can be started, an optimum schedule for the operations can be devised by constructing a graph in which each operation is represented by a point and each point is labeled with a number that represents the time needed for completing the operation. The sequences in which certain operations must be done are indicated by arrowheads on the lines. To determine an optimum schedule the digraph is searched, with a computer if necessary, for a "critical path" that completes the process in a minimum amount of time. Complicated transportation problems can be handled the same way. For example, each line in a digraph can represent a road and can be labeled with the cost of transporting a particular product on it. Clever algorithms can then be applied to find a directed path that minimizes the total cost of shipping the product from one place to another.

Digraphs also serve as playing boards for some unusual board games. Aviezri S. Fraenkel, a mathematician at the Weizmann Institute of Science in Israel, has been the most creative along these lines. (For a good introduction to a class of digraph games Fraenkel calls annihilation games, see "Three Annihilation Games," a paper Fraenkel wrote with Uzi Tassi and Yaacov Yesha for *Mathematics Magazine*, Vol. 51, No. 1, pages 13–17; January 1978.) In 1976 the excellent game Arrows, which Fraenkel developed with Roger B. Eggleton of Northern Illinois University, was marketed in Israel by Or Da Industries and distributed in the U.S. by Leisure Learning Products of Greenwich, CT.

Traffic Jam, another Fraenkel game, is played on the directed graph in Figure 41. A coin is placed on each of four spots: A, D, F, and M. Players take turns moving any one of the coins along one of the lines of the graph to an adjacent spot as is indicated by the arrowheads on the graph. A coin can be moved to any adjacent spot whether or not the spot is occupied, and each spot can hold any number of coins. Note that all the arrowheads at C point inward. Graph theorists call such a point a sink. Conversely, a point from which all the arrowheads point outward is called a source. (If the graph models a pecking order, the sink is the chicken all the other chickens peck and the source is the chicken that pecks all the others.) In this case there is just one sink and

Figure 41

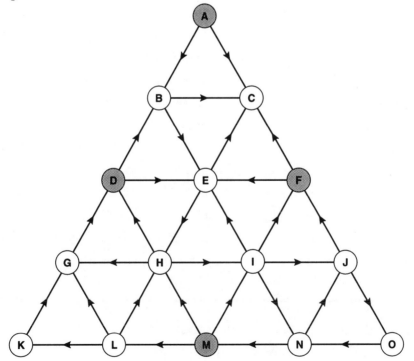

one source. (A complete digraph can never have more than one sink or more than one source. Do you see why?)

When all four coins are on sink C, the person whose turn it is to move has nowhere to go and loses the game. In Conway's book *On Numbers and Games* (Academic Press, 1976) he proves that the first player can always win if and only if his first move is from M to L. Otherwise the opponent can force a win or draw. (It is assumed that both players make their best moves.) With the powerful game theory that Conway has developed it is possible to completely analyze any game of this type, with any starting pattern of counters.

An ancient and fascinating class of puzzles that are best analyzed by digraphs are those known as river-crossing problems. Consider a classic puzzle that turned up in the title of Mary McCarthy's novel *Cannibals and Missionaries.* In the simplest version of this problem three missionaries and three cannibals on the right bank of a river want to get to the left bank by means of a rowboat that can hold no more than two passengers at a time. If the cannibals outnumber the missionaries on either bank, the missionaries will be killed and eaten. Can all six get safely across? If they can, how is it done with the fewest crossings? (I shall not enter here into the current lively debate about whether cannibalism ever actually prevailed in a culture.)

Benjamin L. Schwartz, in an article titled "An Analytic Method for the 'Difficult Crossing' Puzzles" (*Mathematics Magazine,* Vol. 34, No. 4, pages 187-193; March-April 1961), explained how to solve such problems by means of digraphs, but his method deals not directly with the digraphs but rather with their adjacency matrixes. I shall describe here a comparable procedure using the digraphs themselves that was first explained by Robert Fraley, Kenneth L. Cooke and Peter Detrick in their article "Graphical Solution of Difficult Crossing Puzzles" (*Mathematics Magazine,* Vol. 39, No. 3, pages 151-157; May 1966). The paper has been reprinted with additions as Chapter 7 of *Algorithms,*

Graphs and Computers by Cooke, Richard E. Bellman and Jo Ann Lockett (Academic Press, 1970). The following discussion is based on that chapter.

Let m stand for the number of missionaries and c for the number of cannibals, and consider all possible states on the right bank. (It is not necessary to consider states on the left bank as well because any state on the right bank fully determines the state on the left one.) Since m can be equal to 0, 1, 2, or 3, and the same is true for c, there are 4×4, or 16, possible states, which are conveniently represented by the matrix in Figure 42. Six of these states are not acceptable, however, because the cannibals outnumber the missionaries on one of the banks. The ten acceptable states that remain are marked by placing a point inside each of the ten corresponding cells of the matrix.

The next step is to connect these points by lines that show all possible transitions between acceptable states by the transfer of one or two persons to the other side of the river. The result is the undirected graph in Figure 43. This graph is then transformed into a mixed graph by adding arrowheads to show the direction of each transition. The

Figure 42

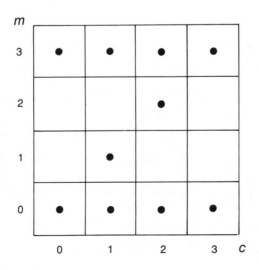

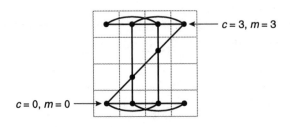

Figure 43

transformation of the undirected graph to a mixed graph must be carried out in accordance with two rules:

1. The object is to create a directed "walk" that will start at the point at the upper right ($c = 3$, $m = 3$) and end at the point at the lower left ($c = 0$, $m = 0$), so that all the cannibals and missionaries end up on the left bank. (This route is called a walk rather than a path because by definition a path cannot visit the same point more than once.)

2. The directed walk must alternate movements down or to the left with movements up or to the right, because each step down or to the left corresponds to a trip from the right bank to the left bank, whereas each step up or to the right corresponds to a trip in the opposite direction.

With both of these rules in mind it takes only a short time to discover that there are just four walks that solve the puzzle. Their digraphs are shown in Figure 44. Each walk completes the transfer in eleven moves. Note that the third through ninth steps are the same in all four walks. The four variants arise because there are two ways to make the first two steps and two symmetrical counterparts for the last two steps.

If the problem is altered to deal with transporting four cannibals and four missionaries (and all other conditions remain the same), the digraph technique can be applied to show there is no solution. Sup-

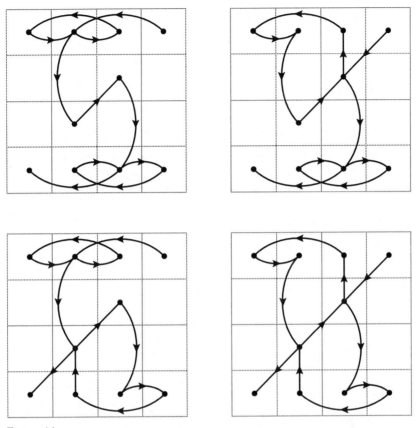

Figure 44

pose now that the boat is enlarged to hold three passengers and that on the boat, as on the bank, the cannibals must not outnumber the missionaries. Under these conditions all eight can cross safely in as few as nine steps. Five cannibals and five missionaries can also cross in a boat that holds three passengers (in eleven steps), but six cannibals and six missionaries cannot.

It is easy to see that given a boat holding four or more passengers any group evenly divided between cannibals and missionaries can be safely transported across the river. One cannibal and one missionary simply do all the rowing, transporting the others one cannibal-mis-

sionary pair at a time until the job is done. Now let n be the number of
cannibals (or missionaries). If the boat holds just four passengers, the
problem is solvable in $2n - 3$ steps. If the boat holds an even number
of passengers that is greater than 4, more than one cannibal–missionary pair can of course be taken each time. The technique of always
keeping the same number of cannibals and missionaries on both sides
of the river is diagrammed as a braided pattern along the diagonal of
the matrix of the problem as is shown in Figure 45. This nine-step
digraph solves the cannibal–missionary problem when n equals 6 and
the boat holds four passengers.

Figure 45

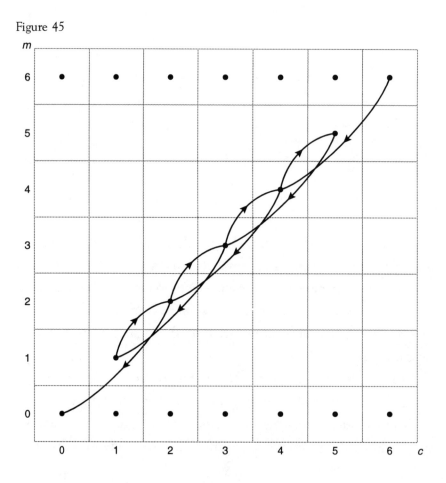

When the capacity of the boat is an even number greater than or equal to 4, the diagonal method always gives the best solution. If the number of cannibals n is just one more than the capacity of the boat, which is an even number greater than 4, then there is always a five-step minimum solution. Actually the diagonal method is more powerful than this last case implies. With a boat that holds an even number greater than 4 it will always provide a five-step minimum solution for any case from $b + 1$ cannibals through $(3b/2) - 2$ cannibals, where b is the capacity of the boat.

If the number of passengers the boat can hold is odd, moving down the diagonal does not always give the best answer. For example, if n equals 6 and the boat holds five, the diagonal method gives the same nine-move solution shown in Figure 45, but the problem also has a seven-step solution. More generally, if the boat holds an odd number of passengers that is greater than three and one less than n, there always is a minimum solution in seven moves. Can you find one of many seven-step solutions for six cannibals and six missionaries crossing the river in a boat that holds five passengers? This is the simplest of an infinity of examples in which, for a boat with an odd capacity, there is a procedure superior to the diagonal procedure. (I am ignoring here the trivial cases of a boat with an odd capacity of one or three, where the diagonal method will not work at all.) The next simplest case is the one where n equals 10 and the boat holds seven passengers.

The digraph method can be applied to almost any kind of river-crossing problem. One famous problem, which goes back at least to the eighth century, concerns three jealous husbands and their wives, who want to cross a river in a boat that holds two passengers. How can this goal be accomplished so that a wife is never alone with a man who is not her husband? If you construct the digraph for the problem, you may be surprised to discover that it is solved by the same four walks as the classic cannibal–missionary problem and has no other solutions.

The only difference—and this applies also to generalizations of the jeal-ous-husband variant of the puzzle—is that the pairings of individual men and women have to be manipulated to meet conditions not essen-tial to the cannibal–missionary version.

Many puzzle books include more exotic variations of the canni-bal–missionary problem. For example, in some cases only certain people may be able to row. (In the classic problem if only one canni-bal and one missionary can row, the solution requires 13 crossings.) The boat may also have a minimum capacity (greater than one) as well as a maximum capacity. Or missionaries may outnumber canni-bals and be safe only if they outnumber them at all times. An island in the river may also be employed as a stopover spot, and certain pairs of individuals may be singled out as being too incompatible to be left alone together.

An ancient problem of this last type (it too can be traced back to the eighth century) is about a man who wants to ferry a wolf, a goat, and a cabbage across a river in a boat that allows him to take only one of them at a time. He cannot leave the wolf alone with the goat or the goat alone with the cabbage. In this case there are two minimal solutions, each of which requires seven trips. One of these solutions is shown in Figure 46, taken from *Moscow Puzzles*, by Boris A. Kordemsky (Charles Scribner's Sons, 1972). Interested readers will find a good selection of such river-crossing problems in books by the British puzzle expert Henry Ernest Dudeney.

I have space for one more digraph puzzle. Paul Erdös has shown that on a complete digraph for n points, when n is less than 7, it is not possible to place arrowheads so that for any two specified points it is always possible to get to each point in one step from some third point. Figure 47 shows a complete graph for seven points. Think of the points as towns joined by one-way roads. Your task is to label each road with an arrowhead so that for any specified pair of towns there is a third

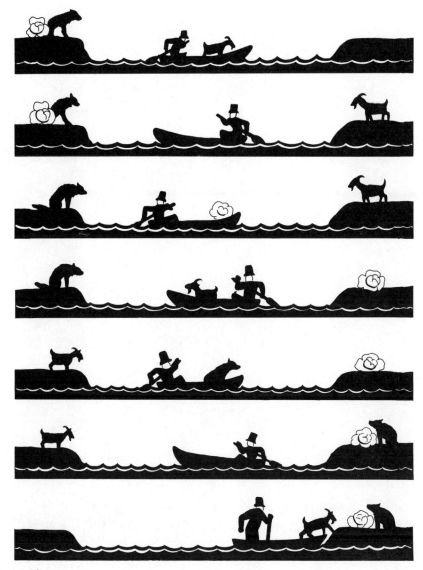

Figure 46

town from which you can drive directly to each of the other two. There is only one solution.

Graphs of this sort are usually called *tournament graphs* because the points can represent players, and the arrows show who beats who. In

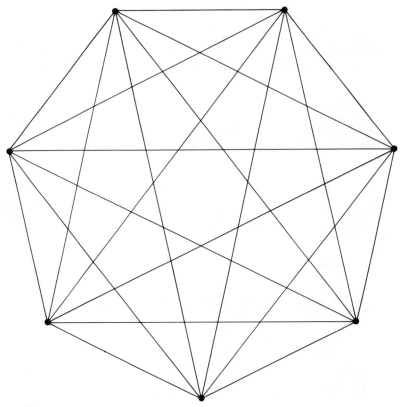

Figure 47

this interpretation, no graph with fewer than seven points can show that for any two players there is always a third person who beats them both. The seven-point graph is the smallest in which this can be the case. It is nontransitive. There is no "best" player because each player can be defeated by another person.

Answers

The unique Hamiltonian circuit is found by starting at A and following a directed path that spells AMBIDEXTROUS. One more step joins S to A, honoring *Scientific American*.

Figure 48 shows a digraph for one of many seven-step solutions to

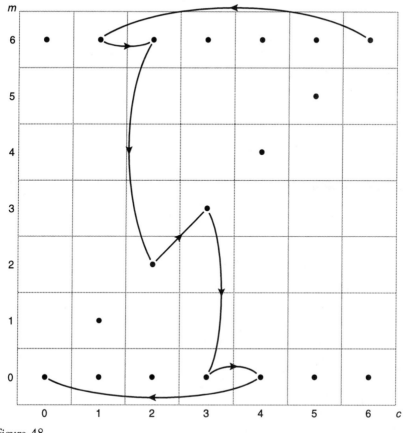

Figure 48

the problem of six missionaries and six cannibals who want to cross a river safely in a boat that holds five.

The Paul Erdös problem is solved by placing arrows on the complete graph for seven points as is shown in Figure 49. Of course, the points and their connecting lines can be permuted in any way to provide solutions that do not appear in this symmetrical form, but all such solutions are topologically the same. See "On a General Problem in Graph Theory," by Paul Erdös in *The Mathematical Gazette* (Vol. 47, No. 361, pages 220–223; October 1963).

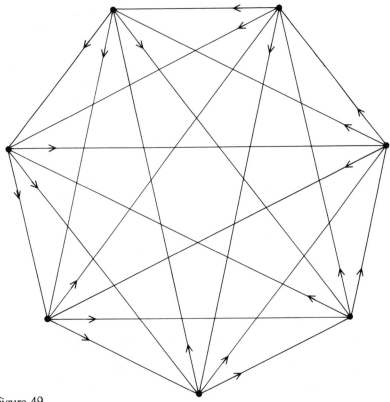

Figure 49

ADDENDUM

Frank Harary was the first to define the distance matrix, the reachability matrix, and the detour matrix, as well as the first to introduce many other graph theory terms that are now standard such as strongly and weakly connected digraphs. This is why Gerhard Ringel, reviewing Harary's classic textbook *Graph Theory*, called him the graph theory Pope. It is because Harary gives the word!

For many years Harary has been inventing and solving two-person games played on graphs. He calls a game in which a defined goal is reached by the winner an "achievement game." If the first person forced to reach the goal is the loser, it is an "avoidance game." His massive work on both types of games remains, alas, unpublished except for occasional papers.

An example of one of Harary's digraph games, which he described to me in a 1980 letter, is a game he calls Kingmaker. Every tournament graph—a complete digraph, every pair of points joined by an arc or directed line—has at least one point called the King that has a distance of 1 or 2 from every other point. This is sometimes known as the King Chicken Theorem.

Kingmaker starts with an undirected complete graph of n points. The first player draws an arrow on any line. Of course it doesn't matter what line he selects because all are alike for symmetry reasons. (Harary suggests that the second player and all onlookers shout "Shrewd move!" after this first arrow is drawn.) The winner is the first to produce a King, in this case a point with a distance of 1 or 2 from all points joined directly from the King by arrows. This usually occurs before all the lines are oriented. In the avoidance game, the player forced to make a King loses. This tends to occur after almost all lines have an arrow.

Steve Maurer, at Swarthmore College, has done much of the work on theorems involving Kings. Every tournament—that is, every complete digraph—must have at least one King, but no such graph can have exactly two Kings. If there are two, there must be a third. Interpreting the points as chickens, a chicken who pecks every other chicken must be the group's only King. A chicken pecked by all the others cannot be a King. A graph with an odd number of points (chickens) can consist entirely of Kings. These theorems provided an amusing page of brain teasers titled "Chicken a la King," by Maxwell Carver (a pseudonym of Joel Spencer), in Discover, March 1988, page 96.

Digraphs furnish a neat, little known method for diagramming problems in the propositional calculus of formal logic. See "The Propositional Calculus with Directed Graphs," on which Harary and I collaborated (giving me my first Erdös number of 2). It appeared in Cambridge University's undergraduate mathematics journal Eureka, March 1988, pages 34-40. The technique is also covered in an appendix added to the second edition (University of Chicago Press, 1982) of my Logic Machines and Diagrams.

References

FINITE GRAPHS AND NETWORKS: AN INTRODUCTION WITH APPLICATIONS. Robert
G. Busacker and Thomas L. Saaty. McGraw-Hill Book Company, 1965.

ONE MORE RIVER TO CROSS. T. H. O'Beirne in *Puzzles and Paradoxes*. Oxford
University Press, 1965.

STRUCTURAL MODELS: AN INTRODUCTION TO THE THEORY OF DIRECTED GRAPHS.
Frank Harary, Robert Z. Norman and Dorwin Cartwright. John Wiley
& Sons, Inc., 1965.

THE THEORY OF ROUND ROBIN TOURNAMENTS. Frank Harary and Leo Moser,
in *American Mathematical Monthly*, Vol. 73, pages 231-246; March 1966.

GRAPH THEORY. Frank Harary. Addison-Wesley, 1969.

TOPICS ON TOURNAMENTS. J. W. Moon. Holt, 1968.

WHEELS WITH WHEELS. Donald E. Knuth in *Journal of Combinatorial Theory*,
Series B, Vol. 16, pages 42-46; 1974. Explains a simple way to represent
strongly connected digraphs.

GRAPHS WITH ONE HAMILTONIAN CIRCUIT. J. Sheehan in *Journal of Graph
Theory*, Vol. 1 pages 37-43; 1977.

ACHIEVEMENT AND AVOIDANCE GAMES FOR GRAPHS. Frank Harary in *Annuals
of Discrete Mathematics*, Vol. 13, pages 111-120; 1982.

KINGMAKER, KINGBREAKER AND OTHER GAMES PLAYED ON A TOURNAMENT. Frank
Harary in the *Journal of Mathematics and Computer Science*, Mathematics
Series, Vol. 1, pages 77-85; 1988.

THE JEALOUS HUSBANDS AND THE MISSIONARIES AND CANNIBALS. Ian Pressman
and David Singmaster in *The Mathematical Gazette*, Vol. 73 pages 73-
81; June 1989.

THE FARMER AND THE GOOSE—A GENERALIZATION. Gerald Gannon and Mario
Martelli in *The Mathematics Teacher*, Vol. 86, pages 202-203; March
1993.

GRAPHS AND DIGRAPHS. Third edition. Gary Chartrand and Linda Lesniak.
Wadsworth, 1996.

8

Dinner Guests, Schoolgirls, and Handcuffed Prisoners

A woman plans to invite 15 friends to dinner. For 35 days she wants to have dinner with exactly three friends a day, and she wants to arrange the triplets so that each pair of friends will come only once. Is this arrangement possible?

That question and others like it, which belong to a vast area of combinatorics called block-design theory, were investigated intensively in the 19th century chiefly as recreational problems. Later they turned out to have an important role in statistics, particularly in the design of scientific experiments. A small branch of block-design theory deals with Steiner triple systems, of which the dinner-guest problem is a simple example. Jacob Steiner, a Swiss geometer, pioneered the study of these systems in the 19th century.

In general a Steiner triple system is an arrangement of n objects in triplets such that each pair of objects appears in a triplet once and only once. It is easy to show that the number of pairs is $1/2n(n-1)$ and that the number of required triplets is one-third the number of pairs, or $1/6n(n-1)$. Of course, a Steiner triple system is possible only when each object is in $1/2(n-1)$ triplets, and these three numbers are integers. That happens when n is congruent to 1 or 3 modulo 6, namely, there is a remainder of 1 or 3 when n is divided by 6. Therefore the sequence of possible values for n is 3, 7, 9, 13, 15, 19, 21, and so on.

With only three guests the dinner problem has a trivial solution: all of them come on the same day. Since Steiner triplets are not ordered, the solution is of course unique. There is also a unique solution for seven guests: (1,2,4), (2,3,5), (3,4,6), (4,5,7), (5,6,1), (6,7,2), and (7,1,3). The order of the triplets and the order of the numbers in each triplet can be altered any way you want without changing the basic pattern. In addition, the numbers can also be exchanged. To understand this point think of each guest as wearing a button with a number painted on it. If two or more guests exchange buttons as they please, the new combination is considered to be the same as the old one.

Similarly, for nine guests there is a unique solution, for 13 guests there are two solutions and for 15 guests it has long been known there are 80 basic solutions. For values of n greater than 15 the number of distinct solutions is not known, although it has been proved there is a solution for every value of n. For $n = 19$ there are hundreds of thousands of solutions.

Let us now complicate the Steiner triple systems a bit to make them more interesting. Suppose the woman decides to invite all 15 friends on each of seven days, seating them three to a table at five tables. She wants each pair of friends to be together at a table only once.

Our new problem is equivalent to one of the most famous puzzles in the history of combinatorial mathematics: Kirkman's schoolgirl problem, named for the Reverend Thomas Penyngton Kirkman, a 19th-century amateur British mathematician who was rector of the church at Croft in Lancashire for more than 50 years. Although he was entirely self-taught in mathematics, his discoveries were so original and diverse that he was elected to the Royal Society. In addition to combinatorics he did significant work on knots, finite groups and quaternions. There is a well-known configuration in projective geometry called Pascal's mystic hexagram (six points on a conic curve joined in all possible ways by straight lines) in which certain intersections are known as Kirkman points.

Kirkman was notorious for his biting sarcasm, which he frequently directed at the philosophy of Herbert Spencer. His parody of Spencer's definition of evolution was often quoted: "A change from a nohowish untalkaboutable all-alikeness, to a somehowish and in-general-talkaboutable not-all-alikeness, by continuous somethingelseifications, and sticktogetherations."

Kirkman first published his schoolgirl problem in 1847 in the *Cambridge and Dublin Mathematics Journal,* Vol. 2, pages 191–204. It appeared again in *The Lady's and Gentleman's Diary for the Year* 1851. Here is how he presented it. Every day of the week a teacher takes 15 schoolgirls on a walk. During the walk the girls are grouped in triplets. Can the teacher construct the triplets so that after the seven walks each pair of girls has walked in the same triplet once and only once?

Any solution to this problem is of course a Steiner triple system, but of the 80 basic solutions for $n = 15$ only seven are basic solutions to the schoolgirl problem. Kirkman designs is the name given to Steiner

triple systems with the extra requirement that the triplets be grouped so
that each group exhausts all the objects.

Again the number of pairs of girls is $1/2n(n-1)$ and the number of
days required for the walks is $1/2(n-1)$. The number of girls must be a
multiple of 3. These values are integers only when n is an odd multiple
of 3. Thus the sequence of possible values is 3, 9, 15, 21, and so on, or
the sequence for the Steiner triple systems with every other number left
out. Does every value in the sequence have a solution? Since the time
Kirkman raised the question a host of papers have been written on
the problem, including many by eminent mathematicians. The case
of $n = 3$ is still trivial. The three girls simply go for a walk. The case of
nine girls in four days has a unique solution:

123	147	159	168
456	258	267	249
789	369	348	357

Like the Steiner triple systems, the numbers in a triplet are not
ordered, and so it does not matter how the numbers are permuted,
how the triplets are arranged within each group, or how the digits are
exchanged with each other. All variations obtained by these permutations are considered to be the same solution.

There are many novel methods, including geometric ones, for constructing Kirkman designs. One of them would have delighted Ramón
Lull, the 13th century Spanish theologian whose Ars magna explored
combinations of symbols with the aid of rotating concentric disks. To
find a solution for $n = 9$ draw a circle and write the digits 1 through 8
around it equally spaced. A cardboard disk of the same size is fastened
to the circle with a pin through both centers. Label the center of the
disk 9. On the disk draw a diameter and two scalene triangles as shown
in Figure 50.

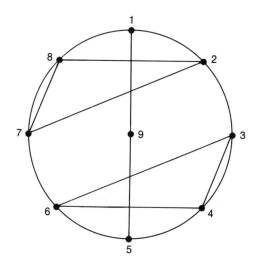

Figure 50

Now rotate the circle in either direction one step at a time to four different positions. (The fifth step brings the pattern back to what it was originally.) At each step copy down the triplet indicated by the ends and the center of the straight line and the two triplets indicated by the corners of the two triangles. The three triplets found at each of the disk's four position give the triplets for each of the four days. This solution seems to be different from the design given above for the schoolgirl problem, but by substituting 2 for 5, 3 for 7, 4 for 9, 5 for 3, 6 for 8, 7 for 6, 8 for 4, and 9 for 2 (and leaving 1 the same) you get the identical design. The only other way to put triangles on the disk to generate a solution is to draw the mirror image of the pattern in the illustration. This procedure, however, will not give rise to a new design.

Since 1922 the case of $n = 15$ has been known to have seven basic solutions. They can be generated by different patterns of triangles, with or without a diameter line. One pattern of five triangles is shown in Figure 51. In this case the disk must be rotated two units at a time to seven different positions. At each position the corners of each triangle provide one of the five triplets for that day.

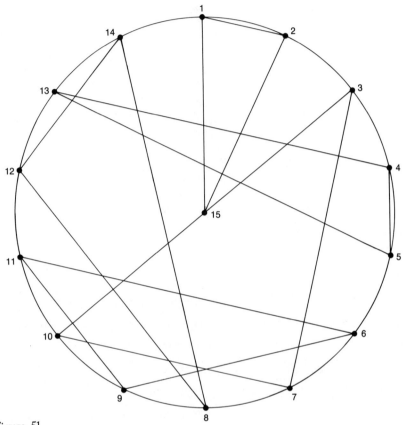

Figure 51

It should be noted that no two triangles on a disk can be congruent. If they were that way, they would duplicate triplets in the overall design. The classic work on Kirkman designs is Chapter 10 of the 11th edition of W. W. Rouse Ball's *Mathematical Recreations & Essays*, revised by H. S. M. Coxeter. The same chapter in the 12th edition of the book (University of Toronto Press, 1974), completely rewritten by J. J. Seidel, is also valuable. The new chapter replaces the early history of the designs with a discussion of how they relate to topics such as affine and projective geometry, Hadamard matrixes, error-correcting codes, Latin squares, and higher-dimensional geometry.

Is there a Kirkman design for every possible value of n? Surprisingly this question went unanswered until 1970, when D. K. Ray-Chaudhuri and Richard M. Wilson of Ohio State University proved that the answer is yes. The number of solutions, however, remains unknown for values of $n = 21$ and all higher values. The proof is presented in "Solution of Kirkman's Schoolgirl Problem" in *Combinatorics* (*Proceedings of Symposia in Pure Mathematics*, Vol. 19, pages 187–203; 1971).

Kirkman designs have many practical uses. Here is a typical way to apply the $n = 9$ design to a biological experiment. Suppose an investigator wants to study the effect of nine environments on a certain animal. There are four species of the animal, and any individual animal can be affected differently depending on whether it is young, fully grown, or aged. Each species is randomly assigned to one of four groups. Within each group are three triplets, each of which includes a randomly picked animal of each age category. Every animal is now assigned to one of the nine environments according to the pattern of nine numbers in its group. This design makes possible an extremely simple way of statistically analyzing the results of the experiment in order to determine what effect the environment has regardless of differences in age and species.

I described above how Kirkman introduced an additional condition that transformed Steiner triple systems into a new kind of block-design problem. In 1917 the British puzzle genius Henry Ernest Dudeney imposed a novel constraint on Kirkman designs that gave rise to still another block-design problem (see Problem 272 of Dudeney's *Amusements in Mathematics* and Problem 287 of his posthumous work *Puzzles and Curious Problems*).

"Once upon a time," begins the second story line of Dudeney's puzzle, "there were nine prisoners of particularly dangerous character who had to be carefully watched. Every week day they were taken out for exercise, handcuffed together, as shown in the sketch made by one of their guards [see Figure 52]. On no day in any one week were the same two men to be handcuffed together. It will be seen how they were sent out on Monday. Can you arrange the nine men in triplets for the remaining five days? It will be seen that No. 1 cannot again be hand-cuffed to No. 2 (on either side), nor No. 2 with No. 3, but, of course, No. 1 and No. 3 can be put together. Therefore it is quite a different problem from the old one of the Fifteen Schoolgirls, and it will be found to be a fascinating teaser and amply repay for the leisure time spent on its solution."

Dudeney gave a solution without explaining how to reach it and

Figure 52

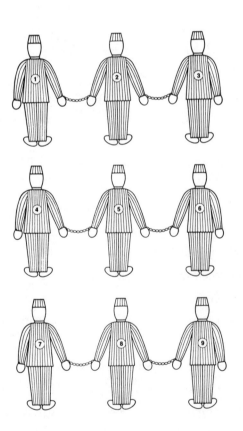

other solutions like it. They can nonetheless be happily found by a Lullian technique with two wheels. A sample pair are shown in Figure 53. Each disk is rotated, say clockwise, three steps at a time. At each step a triplet is generated by the corners of the three triangles. In this case each triplet must have at its center the number indicated by the corner with a spot in it.

Each disk generates the three groups shown below it. In both sets the groups are cyclic in the sense that if you add 3 (modulo 7) to every number in the first group, you get the second group. Similarly, the second group generates the third one, which in turn returns you to the first one. The solution does not start with the pattern given by Dudeney for the first day, although it is easy to exchange digits to obtain that pattern.

After Dudeney answered the puzzle he teased: "If the reader wants a hard puzzle to keep him engrossed during the winter months, let him try to arrange twenty-one prisoners so that they can all walk out, similarly handcuffed in triplets, on fifteen days without any two men being handcuffed together more than once. In case he should come to the opinion that the task is impossible, we will add that we have written out a perfect solution. But it is a hard nut!"

It is a hard nut indeed. As far as I know the first published solution is in Pavol Hell and Alexander Rosa's "Graph Decompositions, Handcuffed Prisoners and Balanced P-Designs" in *Discrete Mathematics* (Vol. 2, No. 3, pages 229–252; June 1972).

Before I give the solution I should like to make a few general remarks about the handcuffed-prisoner problem. The number of pairs of prisoners is $1/2n(n-1)$, as it was with Steiner triple systems and Kirkman designs, although the new restraint (handcuffs!) lengthens the required number of days to $3/4(n-1)$. There is a solution only when this expression is an integer, which is the case when n has a value in a sequence consisting of exactly half of the possible values for a Kirkman design,

Figure 53

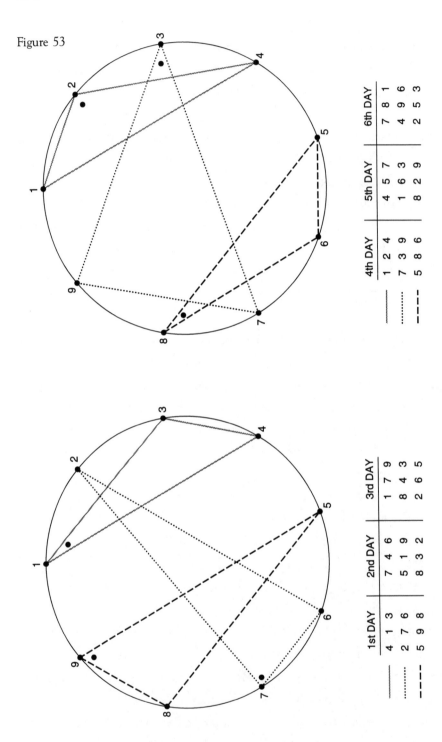

namely the sequence 9, 21, 33, 45, 57, 69, 81, 93, and so on, in which the difference between each adjacent pair of integers is 12.

In 1971 Charlotte Huang and Rosa published a classification of 334 basic solutions for $n = 9$. When Dame Kathleen Ollerenshaw and the cosmologist Hermann Bondi checked each of the solutions, however, they found two duplications among the 334. The actual number of solutions is now thought to be 332. For all values of n greater than 9 the number of solutions is not known. For $n = 21$ Rosa thinks the number of solutions is in the millions. Hell and Rosa have proved that an infinite number of ns have solutions, and they have shown how to find cyclic solutions for all ns less than 100 with the exception of 57, 69, and 93. Wilson (who had helped to crack the Kirkman schoolgirl problem) has demonstrated that all values of n have a solution.

Figure 54 shows a cyclic solution found by Hell and Rosa for $n = 21$. The first seven days form a cyclic set that can be generated by a disk with seven triangles whose corners correspond to the triplets heading each day's design. The disk is rotated three steps at a time. A second disk with seven triangles similarly generates the next seven days, and the 15th day has the design shown at the right in the illustration. In both cyclic sets a day's design can be changed to the next day's by adding 3 (modulo 21) to each number, and doing the same to the last day's design brings back the pattern of the first day. Hell and Rosa give similar cyclic solutions for $n = 33$ and $n = 45$.

Both the schoolgirl problem and the prisoner problem can be generalized to quartets, quintets, sextets, and so on. Such generalization leads to deep combinatorial enigmas, many of which are far from answered. Hundreds of related problems appear in puzzle books, often with story lines about seating arrangements, game tournaments, committee memberships, and other combinatorial schemes. For example, I am often asked how to arrange n members of a bridge club (n must be a multiple of 4) so that they can meet daily for $n - 1$ days at $n/4$ tables

Figure 54

15

1	2	3
4	5	6
7	8	9
10	11	12
13	14	15
16	17	18
19	20	21

7

19	5	15
20	1	17
21	4	12
7	8	3
2	13	18
16	6	14
10	9	11

14

19	1	16
4	13	6
7	20	3
10	14	5
8	11	17
2	9	18
21	15	12

6

16	2	12
17	19	14
18	1	9
4	5	21
20	10	15
13	3	11
7	6	8

13

16	19	13
1	10	3
4	17	21
7	11	2
5	8	14
20	6	15
18	12	9

5

13	20	9
14	16	11
15	19	6
1	2	18
17	7	12
10	21	8
4	3	5

12

13	16	10
19	7	21
1	14	18
4	8	20
2	5	11
17	3	12
15	9	6

4

10	17	6
11	13	8
12	16	3
19	20	15
14	4	9
7	18	5
1	21	2

11

10	13	7
16	4	18
19	11	15
1	5	17
20	2	8
14	21	9
12	6	3

3

7	14	3
8	10	5
9	13	21
16	17	12
11	1	6
4	15	2
19	18	20

10

7	10	4
13	1	15
16	8	12
19	2	14
17	20	5
11	18	6
9	3	21

2

4	11	21
5	7	2
6	10	18
13	14	9
8	19	3
1	12	20
16	15	17

9

4	7	1
10	19	12
13	5	9
16	20	11
14	17	2
8	15	3
6	21	18

1

1	8	18
2	4	20
3	7	15
10	11	6
5	16	21
19	9	17
13	12	14

8

1	4	19
7	16	9
10	2	6
13	17	8
11	14	20
5	12	21
3	18	15

such that each player is the partner of every other player exactly once and the opponent of every other player exactly twice.

The bridge problem seems quite simple, but actually it is so thorny that it was not completely solved until a few years ago. The fullest analysis can be found in "Whist Tournaments," an article by Ronald D. Baker of the University of Delaware. (The article appeared in *Proceedings of the Sixth Southeastern Conference on Combinatorics, Graph Theory and Computing,* published in 1975 by Utilitas Mathematica, Winnipeg, as Volume 14 of the series *Congressus Numerantium.*) Baker shows how to find solutions for all values of n except 132, 152, and 264. Since then the Israeli mathematician Haim Hanani has cracked the case of $n = 132$, and Baker and Wilson have solved the cases of $n = 152$ and $n = 264$.

For many values of n, solutions are generated by disks that rotate one step at a time. Figure 55 shows disks for $n = 4$ and $n = 8$. The technique for generating the solutions is straightforward. A line is drawn from 1 (the disk's center) to 2. Another line is drawn to connect two other numbers. The end points of each line are bridge partners, and the pairs of partners are opponents at the same table. If there is a second table, two more pairs of numbers are joined with colored lines to indicate the seating arrangement for the table. More colors are introduced for additional tables.

The arrangement of such lines generates a cyclic solution if and only if two conditions are met. First, no two lines are the same length (as length is measured by the number of units a line spans on the circumference). Except for the radius line the lengths will necessarily be consecutive integers starting with 1 and ending with $1/2(n - 1)$. Second, if all the opponents at each table are connected by lines (which are broken in the illustration), each length will appear on the disk only twice.

The lines are positioned chiefly by trial and error. No known procedure guarantees a correct pattern for all values of n. Once a pattern is

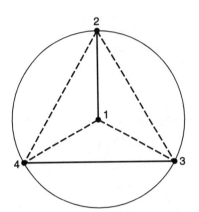

DAYS	TABLE 1	
1	12	34
2	13	42
3	14	23

$n = 4$

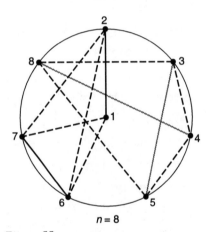

DAYS	TABLE 1		TABLE 2	
1	12	67	35	48
2	13	78	46	52
3	14	82	57	63
4	15	23	68	74
5	16	34	72	85
6	17	45	83	26
7	18	56	24	37

$n = 8$

Figure 55

found it indicates the seating arrangement for the first day. Rotating the disk generates the arrangements for the remaining days. Every column of the final design is cyclic, so that once the seating arrangement for the first day is determined a seating chart for the other days can be rapidly completed without having to turn the disk. The solutions for $n = 132$, $n = 152$, and $n = 264$ are not cyclic, although it may be possible to put them in a cyclic form by permuting the numbers. According to Baker, all values of n may have cyclic solutions, although no general algorithm is known for finding them.

Now for a pleasant problem. Can you design a disk for 12 bridge players that will generate a cyclic tournament meeting all the desired conditions?

Answers

Figure 56 gives two answers to the problem of designing a bridge tournament for 12 players so that they meet at three tables for 11 days and each player is a partner of every other player just once and an opponent of every other player just twice. The first day's distribution is given by a disk on which partners are connected by colored lines and tables are denoted by lines of matching color. Rotating the disk clockwise one step at a time generates the cyclic design for the remaining 10 days. Disk patterns other than the two shown generate additional solutions.

ADDENDUM

Herbert Spencer's definition of evolution, given in his *First Principles*, went like this: "Evolution is a change from an indefinite, incoherent, homogeneity to a definite, coherent, heterogeneity, through continuous differentions

Figure 56

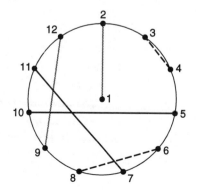 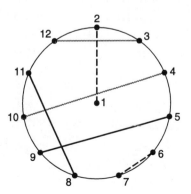

and integrations." Kirkman's travesty appeared in his *Philosophy Without Assumptions* (1876) followed by the question, "Can any man show that my translation is unfair?" Spencer considered it unfair, and in Appendix B to a later edition of *First Principles* he replies at length of what he calls the curious mental states of Kirkman and the mathematician P. G. Tait, who agreed with Kirkman's attacks on evolution.

The solution to the Steiner triple system for seven guests is closely related to a curious solid called the Császár polyhedron. This toroid (it has one hole) is the only known polyhedron, aside from the tetrahedron, that has no diagonals—that is, no lines joining two corners that is not an edge. I describe this solid and show how to make a model in Chapter 11 of *Time Travel and Other Mathematical Bewilderments* (1988).

After this chapter appeared in *Scientific American* in May 1980, I received the following informative letter from Stanford University's computer scientist Donald E. Knuth:

> When I was studying combinatorics in the early 60s it was conventional to say that Steiner triple systems were originally proposed by Steiner in 1853 and solved by Reiss in 1859, with the most elegant known solution being due to E. H. Moore in 1893. But one day I happened to look up a wrong reference to Kirkman's schoolgirl problem and I discovered that Kirkman himself had not only posed the "Steiner" triple problem in 1847, he also solved it elegantly for all n of the form $6k + 1$ and $6k + 3$ and gave the maximum approximate solutions for $6k$ and $6k + 4$. So I told Marshall Hall about this reference, just in time for him to get it into his book *Combinatorial Theory* (1967). Kirkman's schoolgirl problem was the subject of another paper, I think in the same journal and the same year. It is curious how his first paper was apparently forgotten for over 100 years. Perhaps the reason was that he also gave invalid proofs for cases $3k + 2$—his argument for these cases can be paraphrased to say "here is a nice construction and it must be the best possible because God wouldn't want the best answer to be more complicated than this."

I boiled his valid constructions down to less than a page in exercise 6.5-10 of my Volume 3, *The Art of Computer Programming: Sorting and Searching*, they are actually much simpler than Moore's 1893 highly regarded solution to "Steiner's" problem.

References

COMPLETE CLASSIFICATION OF SOLUTIONS TO THE PROBLEM OF 9 PRISONERS. Alexander Rosa and Charlotte Huang in *Proceedings of the 25th Summer Meeting of the Canadian Mathematical Congress*, pages 553-562; June 1971.

GRAPH DECOMPOSITIONS, HANDCUFFED PRISONERS, AND BALANCED P-DESIGNS. Pavol Hell and Alexander Rosa in *Discrete Mathematics*, Vol. 2, pages 229-252; June 1972.

HANDCUFFED DESIGNS. Stephen H. Y. Hung and N. S. Mendelsohn in *Aequationes Mathematicae*, Vol. 11, No. 2/3, pages 256-266; 1974.

ON THE CONSTRUCTION OF HANDCUFFED DESIGNS. J. F. Lawless in *Journal of Combinatorial Theory*, Series A, Vol. 16, pages 74-86; 1974.

FURTHER RESULTS CONCERNING THE EXISTENCE OF HANDCUFFED DESIGNS. J. F. Lawless, in *Aequationes Mathematicae*, Vol. 11, pages 97-106; 1974.

PROJECTIVE SPACE WALK FOR KIRKMAN'S SCHOOLGIRLS. Sister Rita (Cordia) Ehrmann in *Mathematics Teacher*, Vol. 68, No. 1, pages 64-69; January 1975.

KIRKMAN'S SCHOOLGIRLS IN MODERN DRESS. E. J. F. Primrose in *The Mathematical Gazette*, Vol. 60, pages 292-293; December 1976.

HANDCUFFED DESIGNS. S. H. Y. Hung and N. S. Mendelsohn in *Discrete Mathematics*, Vol. 18, pages 23-33; 1977.

THE NINE PRISONERS PROBLEM. Dame Kathleen Ollerenshaw and Sir Hermann Bondi in *Bulletin of the Institute of Mathematics and Its Applications*, Vol. 14, No. 5-6, pages 121-143; May/June 1978.

NEW UNIQUENESS PROOFS FOR THE (5, 8, 24), (5, 6, 12) AND RELATED STEINER SYSTEMS. Deborah J. Bergstrand in *Journal of Combinatorial Theory*, Series A, Vol. 33, pages 247-272; November 1982.

DECOMPOSITION OF A COMPLETE MULTIGRAPH INTO SIMPLE PATHS: NONBALANCED
   HANDCUFFED DESIGNS. Michael Tarsi in *Journal of Combinatorial Theory*,
   Series A, Vol. 34, pages 60–70; January 1983.
GENERALIZED HANDCUFFED DESIGNS. Francis Maurin in *Journal of Combinatorial Theory*, Series A, Vol. 46, pages 175–182; November 1987.

9

The
Monster
and Other
Sporadic
Groups

What's purple and commutes?
An Abelian grape.

—Anonymous mathematical riddle,
                    ca. 1965

All over the world during the last half of the 1970s experts in a branch of abstract algebra called group theory struggled to capture a group that John Horton Conway nicknamed "The Monster." The name derives from its size. When it was finally constructed in 1980, the number of its elements proved to be

808,017,
424,794,512,875,886,459,904,961,710,

$$757{,}005{,}754{,}368{,}000{,}000{,}000, \text{ or}$$
$$2^{46} \times 3^{20} \times 5^9 \times 7^6 \times 11^2 \times 13^3 \times 17 \times 19 \times$$
$$23 \times 29 \times 31 \times 41 \times 47 \times 59 \times 71$$

The man who captured the beast bare-handed was Robert L. Griess, Jr., a mathematician then at the University of Michigan. (His last name rhymes with rice.) Griess dislikes the term monster, preferring to call it "The Friendly Giant" or to refer to it by its mathematical symbol F_1. The news of his discovery was enormously exciting to group theorists because it brought them closer to completing a task that occupied them for more than a century: the classification of all finite simple groups.

The colorful story of this undertaking begins with a bang. In 1832 Évariste Galois, a French mathematical genius and student radical, was killed by a pistol shot in an idiotic duel over a woman. He was not yet 21. Some early, fragmentary work had already been done on groups, but it was Galois who laid the foundations of modern group theory and even named it, all in a long, sad letter that he wrote to a friend the night before his fatal duel.

What is a group? Roughly speaking, it is a set of operations performed on something, with the property that if any operation in the set is followed by any operation in the set, the outcome can also be reached by a single operation in the set. The operations are called the elements of the group, and their number is called the order of the group.

Before going on to a more precise definition let us consider an example. You are standing at attention and must carry out any of four commands: "Do nothing," "Left face," "About face" and "Right face." Now suppose you execute a left face followed by an about face. A sequence of this kind will be called a multiplication of the two operations. Note that the "product" of this particular multiplication can be reached by the single operation right face. This set of four operations is a group because it meets the following axioms.

1. *Closure:* The product of any pair of operations is equivalent to a single operation in the set.

2. *Associativity:* If the product of any two operations is followed by any operations, the result is the same as following the first operation with the product of the second and the third.

3. *Identity:* There is just one operation that has no effect, in this case doing nothing.

4. *Inverse:* For every operation there is an inverse operation such that executing an operation and then its inverse is equivalent to executing the identity operation. In this example left face and right face are inverses of each other, whereas do nothing (the identity) and about face are each their own inverse.

Any set of operations that satisfies these four axioms is a group, and the group of four commands I have just described is called the cyclic 4-group because it can also be modeled by the cyclic permutations of four objects in a row. (In a cyclic permutation of a set of ordered elements the first element moves into the second position, the second element moves into the third position and so on, with the last element moving to the first position.) Label the four objects 1, 2, 3, and 4 and assume that they are lined up in numerical order: 1234. The identity operation, which I shall call I, leaves the order of the objects unaltered. Operation A permutes them to 4123, B to 3412, and C to 2341. This group can be completely characterized by the "multiplication" table at the top right in Figure 57. Each cell in the table gives the operation that is equivalent to performing the operation indicated at the left end of its row followed by the operation indicated at the top of its column. If a similar construction is carried out for the first model, letting I, A, B, and C stand for the four commands ("Do nothing," "Left face," and so on) in their listed order, the same table results, proving that the cyclic 4-group and the group of four commands are isomorphic, or equivalent.

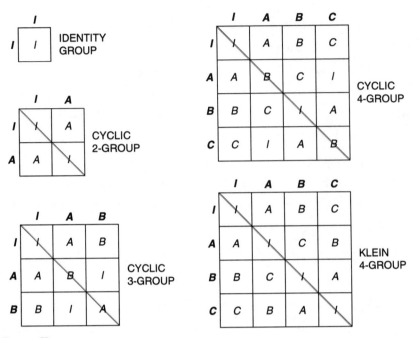

Figure 57

Note that the table for the cyclic 4-group is symmetrical about one of its diagonal axes. This characteristic of the table indicates that the group also obeys the commutative axiom, meaning that the product of any two operations is the same regardless of which one is performed first. Groups that display this property are called Abelian groups after the Norwegian mathematician Niels Henrik Abel. Any cyclic permutation of n objects generates an Abelian group, which is equivalent to the group of the orientation-preserving rotations of a regular polygon of n sides. (A rotation preserves a figure's orientation if the figure ends up in exactly the same position in which it began.) Thus the cyclic 4-group can be modeled by the orientation-preserving rotations of a square.

There is just one group of order 1: the trivial group consisting of the identity operation. It is not hard to see that this operation meets all four of the criteria defining a group. For example, if you do nothing to something twice in succession, it is the same as doing nothing, and so the closure axiom is satisfied. The only order-2 group is almost as trivial. This group, the table for which is shown in Figure 57, can be modeled with two operations to be performed on a penny: doing nothing to the penny (*I*) and turning the penny over (*A*). The only group of order 3 is the cyclic 3-group, which is equivalent to the set of cyclic permutations of three objects and to the set of orientation-preserving rotations of an equilateral triangle. There are just two groups of order 4: the cyclic 4-group and another group known as the Klein 4-group.

The Klein 4-group can be easily modeled with the following operations on two pennies placed side by side: doing nothing (*I*), turning over the left penny (*A*), turning over the right penny (*B*), and turning over both pennies (*C*). The table for the group, shown in Figure 57 reveals that this group too is Abelian.

The simplest example of a non-Abelian group is the set of six symmetry operations on the equilateral triangle: the identity, rotating the triangle 120 degrees clockwise, rotating it 120 degrees counterclockwise, and flipping it over about any of its three altitudes. To prove that the elements of this group do not commute label the corners of a cardboard triangle, rotate the triangle 120 degrees in either direction and turn it over about any altitude; then perform the same two operations in the reverse order and compare the results. If each vertex of the triangle is identified with a different object, the resulting 6-group is equivalent to the group of all permutations on three objects.

To test your understanding of a group you might pause to consider the following three models.

1. With a deck consisting of four face-down playing cards the following operations are defined: the identity operation (*I*), transposing the top two cards in the deck (*A*), transposing the bottom two cards (*B*),

and removing the middle two cards and putting the lower one on the bottom of the deck and the other one on the top (C).

2. A dollar bill is placed either face up or face down and either right-side up or upside down. The operations are the identity (I), rotating the bill 180 degrees (A), turning it over about its vertical axis (B), and turning it over about its horizontal axis (C).

3. A sock is on either the left foot or the right foot in one of two states, right-side out or inside out. The operations are the identity (I); taking off the sock, reversing it, and putting it back on the same foot (A); moving the sock to the other foot without reversing it (B); and taking off the sock, reversing it, and moving it to the other foot (C).

For each of these groups make a multiplication table and determine whether the group is equivalent to the cyclic 4-group or the Klein 4-group.

The multiplication table of a group can be represented graphically by a diagram called a Cayley color graph after the mathematician Arthur Cayley. For example, the graph at the lower left in Figure 58 is a Cayley color graph for the cyclic 4-group, the table for which is at the top of the illustration. The four points of the graph correspond to the four operations of the group. Every pair of points has been joined by a pair of lines going in opposite directions, with the direction of each line indicated by an arrowhead, and a color has been assigned to each operation in accordance with the key shown at the top of the table. To understand how the graph reproduces the information in the table consider the line from B to A. The color of the line is determined by starting at B on the left side of the table, moving to the right to the cell containing A and then using the color assigned to the letter, C, at the top of the column the cell is in. The same procedure yields the colors of all the other lines.

When two points in a Cayley color graph are joined by two lines

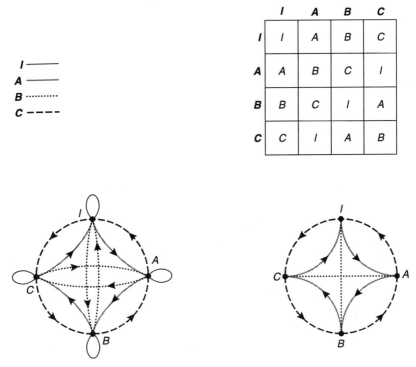

Figure 58

of differing color, the operations represented by the two colors are the inverse of each other. If both lines are the same color, the operation associated with that color is its own inverse. In this case the graph can be simplified by replacing the two directed lines of like color with a single undirected line of that color. In addition the identity operation is represented by a loop that joins each point to itself, and so because one of these loops is at each corner of the graph they can all be omitted. The simplified version of the graph is at the lower right in Figure 58.

A simplified Cayley graph for the Klein 4-group is shown in Figure 59, and one for the non-Abelian permutation 6-group is shown in Figure 60. For graphs of higher order it is more convenient to stop

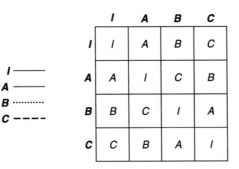

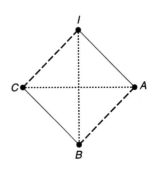

Figure 59

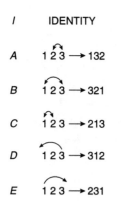

	I	A	B	C	D	E
I	I	A	B	C	D	E
A	A	I	D	E	B	C
B	B	E	I	D	C	A
C	C	D	E	I	A	B
D	D	C	A	B	E	I
E	E	B	C	A	I	D

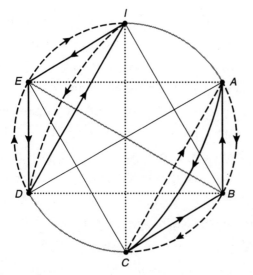

Figure 60

using colors and instead label each line with the symbol corresponding to the operation it represents.

It should be obvious that given the color graph for any group, the table for the group can be constructed. The converse is also true. The graphs are valuable aids, however, because they often reveal properties that are not easily seen in a multiplication table. For example, it is not difficult to see that if on the 6-graph the lines corresponding to operations A, B, and C are omitted, leaving only the lines corresponding to D and E, two disconnected graphs are obtained. Each is a color graph for a cyclic 3-group, but only the set of operations I, D, E actually forms such a group because this set alone contains the identity operation. Any subset of the elements of a group that itself forms a group is called a subgroup, and so inspection of the color graph has revealed that the cyclic 3-group is a subgroup of the permutation 6-group.

So far only groups with a finite number of operations, or elements, have been discussed. There are also infinite groups. They fall into two classes: discrete groups, which have a countable infinity of elements, and continuous groups, which have an uncountable infinity of elements. An infinite set is said to be countable if its members can be matched up one for one with the positive integers 1, 2. . . . Hence the integers themselves are an example of a countably infinite set, whereas the points on the real-number line are an example of an uncountably infinite set. In fact, the integers form a discrete Abelian group under the operation of addition, with 0 as the identity element and $-a$ as the inverse of any element a. The real numbers, on the other hand, form a continuous group with respect to addition, and if 0 is excluded, they form a continuous group with respect to multiplication as well. (In the latter case 1 is the identity element, and $1/a$ is the inverse of a.) Continuous groups are called Lie (pronounced lee) groups after the Norwegian mathematician Marius Sophus Lie. A trivial geometrical example of a Lie group is the group of symmetry rotations of a circle (or a sphere or a hypersphere). The degree of rotation can be as small as one likes.

The group is one of the most powerful and unifying of all concepts in mathematics. Moreover, in addition to turning up in every branch of mathematics, groups have endless applications in science. Wherever there is symmetry there is a group. The Lorentz transformations of relativity theory form a Lie group based on the continuous rotation of an object in space-time. Finite groups underlie the structure of all crystals and are indispensable in chemistry, quantum mechanics, and particle physics. The famous eightfold way, which classifies the family of subatomic particles known as hadrons, is a Lie group. Every geometry can be defined as the study of the properties of a figure that are left invariant by a group of transformations.

Even recreational mathematics sometimes involves groups. Since every finite group can be modeled by a set of permutations on n objects, it is hardly surprising to find groups related to card shuffling, ball juggling, campanology (bell ringing), sliding-block puzzles, and all kinds of combinatorial puzzles such as Rubik's cube. In an earlier column (reprinted in my *New Mathematical Diversions from Scientific American*) I explained how groups apply to braiding theory and so underlie numerous magic tricks involving ropes and twisted handkerchiefs.

In view of the great elegance and utility of groups it is understandable that mathematicians would like to be able to classify them. The Lie groups have been classified, but there are other infinite groups that still elude classification. What about finite groups? One might suppose they would be easier to classify than Lie groups, but that has not proved to be the case. It is this difficult task that is now on the brink of being completed.

All finite groups are constructed from building blocks called simple groups in much the same way that chemical compounds are constructed from elements, proteins from amino acids, and composite numbers from primes. A simple group is one that has no "normal" subgroups other than itself and the trivial identity subgroup. Remember that a subgroup is defined as any subset of the elements of a group that is itself a group. What is meant by "normal" is best explained as follows. Consider a group G with a subgroup S. The set of products obtained by multiplying an element g of the group G by each element of the subgroup S is called a right coset of gS. Similarly, the set of products obtained by multiplying each element of S by g is called a left subset. If for all choices of g we have gS = Sg, that is, if the left and right cosets are equal, then the subgroup is called normal.

For example, the cyclic 3-group is a normal subgroup of the permutation 6-group. Hence the 6-group is not simple. Simple groups are the building blocks of all groups, and so to classify the finite groups it is necessary to classify all the finite simple groups.

Almost all finite simple groups belong to families with an infinity of members. Families of this type provide a quite satisfactory system of classification, since there are procedures for constructing any individual member, or group. For example, the cyclic permutation groups of prime order (which are modeled by rotations of regular polygons with a prime number of sides) are finite simple groups. In fact, they are the only finite simple groups that are Abelian as well as the only ones that are cyclic. A famous result in mathematics called Lagrange's theorem states that the order, or number of elements, of any subgroup must be a divisor of the order of the group in which it is contained. Since a prime number has no divisor (other than 1 and itself), this theorem implies that any group of prime order has no subgroups (other than the identity and itself). If a group has no other subgroups, however, then it certainly has no normal subgroups, and so it follows that any group of prime order is simple.

Another important family of finite simple groups is the set of alternating groups, which are modeled by the even permutations on

n objects for all integers n greater than 4. An even permutation is one that can be obtained in an even number of steps, where each step consists in switching two objects. For example, the cyclic 3-group is also an alternating group because 231 can be produced from 123 in two steps (transpose first 1 and 2 and then 1 and 3) and the same is true for any other pair of the three cyclic permutations of three objects. Exactly half of all permutations are even, and because n objects can be permuted in $n!$ ways every alternating group has an order of $n!/2$. The odd permutations do not form groups because any odd permutation followed by another odd permutation is equivalent to an even permutation, so that the closure axiom is not satisfied.

There are 16 other infinite families of finite simple groups, all of them non-Abelian and noncyclic. The orders of the simple groups (excluding the cyclics) form an infinite sequence that starts with 60, the order of the alternating group on five objects. (This group is equivalent to the group of rotations of a regular dodecahedron or icosahedron.) The sequence begins 60, 168, 360, 504, 660, 1092, 2448, 2520, 3420, 4080, 5616, 6048, 6072, 7800, 7920. . . . If 1 and all the prime numbers are inserted into this infinite sequence, the resulting sequence gives the orders of all finite simple groups.

Unfortunately the list includes a small number of groups (starting with the group of order 7920) that cannot be fitted into any infinite family. These are the non-Abelian anomalies, the jokers that defy all classification. Mathematicians know them as the sporadic simple groups, but they are quite complicated. If there is an infinite number of these sporadics, and if there is no pattern ordering them, then the task of classifying all finite simple groups, and therefore all finite groups, is hopeless. There are, however, compelling reasons for thinking there are no sporadics other than the 26 already identified. (A classic history of sporadic groups appears in Figure 61. First published in The American Mathematical Monthly in November 1973, the ballad is said to have

Figure 61

A Simple Ballad
(To be sung to the tune of "Sweet Betsy from Pike")

What are the orders of all simple groups?
I speak of the honest ones, not of the loops.
It seems that old Burnside their orders has
    guessed
Except for the cyclic ones, even the rest.

Groups made up with permutes will produce
    some more:
For A_n is simple, if n exceeds 4.
Then, there was Sir Matthew who came into
    view
Exhibiting groups of an order quite new.

Still others have come on to study this thing.
Of Artin and Chevalley now we shall sing.
With matrices finite they made quite a list.
The question is: Could there be others they've
    missed?

Suzuki and Ree then maintained it's the case
That these methods had not reached the end of
    the chase.
They wrote down some matrices, just four by
    four,
That made up a simple group. Why not make
    more?

And then came the opus of Thompson and Feit
Which shed on the problem remarkable light.
A group, when the order won't factor by two,
Is cyclic or solvable. That's what is true.

Suzuki and Ree had caused eyebrows to raise,
But the theoreticians they just couldn't faze.
Their groups were not new: if you added a twist,
You could get them from old ones with a flick
    of the wrist.

Still, some hardy souls felt a thorn in their side.
For the five groups of Mathieu all reason defied;
Not A_n, not twisted, and not Chevalley,
They called them sporadic and filed them away.

Are Mathieu groups creatures of heaven or hell?
Zvonimir Janko determined to tell.
He found out what nobody wanted to know:
The masters had missed 1 7 5 5 6 0.

The floodgates were opened! New groups were
    the rage!
(And twelve or more sprouted, to greet the new
    age.)
By Janko and Conway and Fischer and Held,
McLaughlin, Suzuki, and Higman, and Sims.

No doubt you noted the last lines don't rhyme.
Well, that is, quite simply, a sign of the time.
There's chaos, not order, among simple groups;
And maybe we'd better go back to the loops.

been "found scrawled on a library table in Eckhart Library at the University of Chicago; author unknown or in hiding." The "loops" referred to therein are the simple cyclic groups, and A_n is the symbol for the alternating group for n objects.)

The search for the sporadic simple groups began in the 1860s when the French mathematician Émile Léonard Mathieu discovered the first five. The smallest of them, known as M_{11}, has 7,920 operations and is modeled by certain permutations on 11 objects. A century slipped by before the sixth sporadic, of order 175,560, was found in 1965 by Zvonimir Janko of the University of Heidelberg. Three years later John Horton Conway, then at the University of Cambridge, surprised everyone by finding three more sporadics. His work was based on Leech's lattice, a scheme devised by John Leech, a British mathematician, for packing unit hyperspheres densely in a space of 24 dimensions. (In Leech's lattice each hypersphere touches exactly 196,560 others.)

Leech discovered his lattice while working on error-correcting codes. It turns out that there is a close connection between certain sporadic groups and codes employed in reconstructing a message distorted by noise. Two of Mathieu's sporadic groups, M_{23} and M_{24}, are related to the Golay error-correcting code that is often used for military purposes. Roughly speaking, a good error-correcting code is based on a subset of unit hyperspheres placed as for apart from one another as is possible in a dense packing.

At the start of 1980 two dozen sporadic groups had been proved to exist, and two more, J_4 and F_1, were believed to be authentic. (A complete list of these 26 sporadic groups is shown in Figure 62.) J_4, which was proposed by Janko in 1975, was finally constructed in February by David Benson, Conway, Simon P. Norton, Richard Parker, and Jonathan Thackray, a group of mathematicians at Cambridge. F_1 (the monster), which is by far the largest sporadic, was conjectured independently by Griess and by Bernd Fischer of the University of Bielefeld in 1973 and

Figure 62

NAME OF GROUP	NUMBER OF ELEMENTS	DISCOVERED BY
M_{11}	$2^4 \times 3^2 \times 5 \times 11$	Mathieu
M_{12}	$2^6 \times 3^3 \times 5 \times 11$	
M_{22}	$2^7 \times 3^2 \times 5 \times 7 \times 11$	
M_{23}	$2^7 \times 3^2 \times 5 \times 7 \times 11 \times 23$	
M_{24}	$2^{10} \times 3^3 \times 5 \times 7 \times 11 \times 23$	
J_1	$2^3 \times 3 \times 5 \times 7 \times 11 \times 19$	Janko
J_2	$2^7 \times 3^3 \times 5^2 \times 7$	Hall, Wales
J_3	$2^7 \times 3^5 \times 5 \times 17 \times 19$	Higman, McKay
J_4	$2^{21} \times 3^3 \times 5 \times 7 \times 11^3 \times 23 \times 29 \times 31 \times 37 \times 43$	Benson, Conway, Janko, Norton, Parker, Thackray
HS	$2^9 \times 3^2 \times 5^3 \times 7 \times 11$	Higman, Sims
MC	$2^7 \times 3^6 \times 5^3 \times 7 \times 11$	McLaughlin
Sz	$2^{13} \times 3^7 \times 5^2 \times 7 \times 11 \times 13$	Suzuki
C_1	$2^{21} \times 3^9 \times 5^4 \times 7^2 \times 11 \times 13 \times 23$	Conway
C_2	$2^{18} \times 3^6 \times 5^3 \times 7 \times 11 \times 23$	
C_3	$2^{10} \times 3^7 \times 5^3 \times 7 \times 11 \times 23$	
He	$2^{10} \times 3^3 \times 5^2 \times 7^3 \times 17$	Held, Higman, McKay
F_{22}	$2^{17} \times 3^9 \times 5^2 \times 7 \times 11 \times 13$	Fischer
F_{23}	$2^{18} \times 3^{13} \times 5^2 \times 7 \times 11 \times 13 \times 17 \times 23$	
F_{24}	$2^{21} \times 3^{16} \times 5^2 \times 7^3 \times 11 \times 13 \times 17 \times 23 \times 29$	
Ly	$2^8 \times 3^7 \times 5^6 \times 7 \times 11 \times 31 \times 37 \times 67$	Lyons, Sims
O	$2^9 \times 3^4 \times 5 \times 7^3 \times 11 \times 19 \times 31$	O'Nan, Sims
R	$2^{14} \times 3^3 \times 5^3 \times 7 \times 13 \times 29$	Conway, Rudvalis, Wales
F_5	$2^{14} \times 3^6 \times 5^6 \times 7 \times 11 \times 19$	Conway, Fischer, Harada, Norton, Smith
F_3	$2^{15} \times 3^{10} \times 5^3 \times 7^2 \times 13 \times 19 \times 31$	Smith, Thompson
F_2	$2^{41} \times 3^{13} \times 5^6 \times 7^2 \times 11 \times 13 \times 17 \times 19 \times 23 \times 31 \times 47$	Fischer, Leon, Sims
F_1	$2^{46} \times 3^{20} \times 5^9 \times 7^6 \times 11^2 \times 13^3 \times 17 \times 19 \times 23 \times 29 \times 31 \times 41 \times 47 \times 59 \times 71$	Fischer, Griess

constructed in January by Griess, as I have mentioned. Several much smaller sporadics, the construction of which required long computer calculations, are embedded in F_1 in such a way that their existence follows almost trivially from the existence of F_1. Yet to everyone's astonishment Griess's construction of F_1 was carried out entirely by hand. F_1 is said to be based on a group of symmetry rotations in a space of 196,883 dimensions!

Is the list of 26 sporadics complete? Most group theorists are convinced it is, but the task of proving the conjecture could be formidable. Indeed, the final published proof is likely to require as many as 10,000 printed pages. It should be noted, however, that proofs in group theory tend to be unusually long. A famous proof by John Thompson and Walter Feit, which among other things established William Burnside's conjecture that all non-Abelian finite simple groups are of even order, covered more than 250 pages: an entire issue of *The Pacific Journal of Mathematics* (Vol. 13, pages 775–1029; 1963).

In 1972 Daniel Gorenstein of Rutgers University outlined a 16-step program for completing the classification of the finite simple groups. This guide to a final proof was soon improved and greatly "speeded up" by Michael Aschbacher of the California Institute of Technology. Both men are world experts on groups. (Aschbacher later won the much coveted Cole Prize in algebra.) In May 1977, Gorenstein told *The New York Times* that he had been working on the classification problem five hours a day, seven days a week, 52 weeks a year since 1959. "I want to solve it," he said, "because I want to solve it, not because it will benefit mankind." Like most other group theorists, Gorenstein is convinced that no new sporadic groups will be found, and that a proof that the list of 26 groups is almost complete.

There is, of course, no way to predict whether a practical application will or will not be found for any mathematical result whose discovery was not motivated by practical considerations. We do know that groups lie at the very heart of the structure of the universe. Nature seems to prefer small, uncomplicated groups, but this could be an illusion created by the fact that applications of small groups are the easiest to find, particularly in a world limited to three spatial dimensions. Who can say that at some far distant date, if the human race survives, even the monster will not turn out to have some remarkable but at present unimaginable application?

Answers

The problem was to determine whether three models of a 4-group are examples of the cyclic 4-group or the Klein 4-group. All three are Klein 4-groups. An easy test for determining the character of a given 4-group is checking to see whether each operation in the group is its own inverse. If it is, the group is a Klein 4-group.

ADDENDUM

The final classification of all finite simple groups was completed in August 1980. Known as "The Enormous Theorem," the proof rests on hundreds of papers by more than a hundred mathematicians around the world, written over the last three decades. The complete proof, when published in many volumes, is expected to require some 5,000 pages! Whether it can be simplified and shortened remains to be seen. As everyone had anticipated, there are exactly 26 sporadic groups.

"Simple groups are beautiful things," wrote John Conway shortly before the Enormous Theorem was proved, "and I'd like to see more of them, but am reluctantly coming around to the view that there are likely no more to be seen."

When the four-color map theorem was established by a horrendous computer printout, some mathematicians suggested that computers were introducing a qualitatively different kind of proof. Because computers are machines that can make mistakes, it was argued, such proofs support the view that mathematics is an empirical science, as fallible as physics. The Enormous Theorem belies this curious view. It is far more massive than the map theorem printout and just as prone to error if not more so. Indeed, one can claim that computer proofs are more accurate than enormously long hand proofs because a computer proof can be reprogrammed a different way, and one program used to check the other. Moreover, a computer differs from an adding machine, or even an abacus, only in the speed with which it manipulates symbols. A mathematician who uses a modern computer is in no essential way different from a mathematician who uses a hand calculator to do large multiplications and divisions.

Physicist Tony Rothman, in "Genius and Biographers: The Fictionalizations of Évariste Galois" (*American Mathematical Monthly*, Vol. 89, February 1982) and "The Short Life of Évariste Galois" (*Scientific American*, April 1982), presented evidence that Eric Temple Bell, in his popular *Men of Mathematics* (1937), overromanticized many facts. In Bell's account, Galois spent the night before his duel "feverishly dashing off" what he had discovered about groups. "Time after time," wrote Bell, "he broke off to scribble in the margin 'I have not time. I have not time.'"

Although young Galois was indeed killed in a duel over a woman, this passage is almost entirely wrong. Galois had written several articles on group theory, and was merely annotating and correcting those earlier published papers. "There are a few things left to be completed in this proof," he wrote in the margin. "I have not time." That was the sole basis for Bell's statement about Galois writing over and over "I have not time. I have not time."

References

THE THEORY OF GROUPS. Marshall Hall, Jr. Macmillan, 1959.

FINITE GROUPS. Daniel Gorenstein. Harper and Row, 1968.

THE FASCINATION OF GROUPS. F. J. Budden. Cambridge, 1972.

THE SEARCH FOR FINITE SIMPLE GROUPS. Joseph A. Gallian in *Mathematics Magazine*. Vol. 49, No. 4, pages 163–180; September 1976.

GROUPS AND SYMMETRY. Jonathan L. Alperin in *Mathematics Today*, edited by Lynn Arthur Steen. Springer-Verlag, 1978.

A MONSTROUS PIECE OF RESEARCH. Lynn Arthur Steen in *Science News*, Vol. 118, pages 204–206; September 27, 1980.

THE FINITE SIMPLE GROUPS AND THEIR CLASSIFICATION. Michael Aschbacher. Yale, 1980.

MONSTERS AND MOONSHINE. John Conway in *The Mathematical Intelligencer*, Vol. 2, pages 165–171; 1980.

FINITE SIMPLE GROUPS. Daniel Gorenstein. Plenum, 1982.

THE CLASSIFICATION OF FINITE SIMPLE GROUPS, Vols. 1 and 2. Daniel Gorenstein. Plenum, 1983.

ATLAS OF FINITE GROUPS. R. T. Curtis, S. P. Norton, R. A. Parker, and R. A. Wilson. Clarendon, 1985.

THE ENORMOUS THEOREM. Daniel Gorenstein in *Scientific American*, pages 104–115; December 1985.

TEN THOUSAND PAGES TO PROVE SIMPLICITY. Mark Cartwright in *New Scientist*, Vol. 106, pages 26–30; May 30, 1985.

DEMYSTIFYING THE MONSTER. Ian Stewart in *Nature*, Vol. 319, pages 621–622; February 20, 1986.

ARE GROUP THEORISTS SIMPLEMINDED? Barry Cipra in *What's Happening in the Mathematical Sciences*, Vol. 3. American Mathematical Society, 1996.

A HUNDRED YEARS OF FINITE GROUP THEORY. Peter M. Neumann in *The Mathematical Gazette*, pages 106–118; March 1996.

10

Taxicab Geometry

A conjecture both deep and
    profound
Is whether a circle is round.
  In a paper by Erdös,
  Written in Kurdish,
A counterexample is found.

—Anonymous

Altering one or more postulates of Euclidean geometry makes it possible to construct all kinds of strange geometries that are just as consistent, or free of internal contradictions, as the plane geometry taught in secondary schools. Some of these non-Euclidean geometries have turned out to be enormously useful in modern physics and cosmology, but the two most important, elliptic geometry and hyperbolic geometry, have a structure that is impossible to visualize. Hence most laymen find these geometries too difficult to comprehend and are

159

certainly not able to search their structure for new theorems or to work on interesting non-Euclidean problems.

In this chapter we shall take an elementary look at a quite different kind of non-Euclidean geometry, one so easy to understand that anyone exploring its structure on ordinary graph paper can have the excitement of discovering new theorems. Often called taxicab geometry, this system can be modeled by taxicabs roaming a city whose streets form a lattice of unit-square blocks. In many ways taxicab geometry is curiously like ordinary plane geometry. Yet it is sufficiently different that exploring it can be great fun. Moreover, such exploration provides a strong feeling for how geometries may vary in bizarre ways from Euclidean geometry and still form a logically consistent formal system.

As far as I know taxicab geometry was first seriously proposed by Hermann Minkowski, a mathematician born in Russia who was young Albert Einstein's teacher in Zurich. Minkowski later gave special relativity its beautiful formulation in a four-dimensional geometry of space and time, and the space–time graphs widely used in relativity theory are named for him. At about the turn of the century he published in Germany his *Collected Works* (reprinted in the U.S. by Chelsea Publishing Company in 1967), in which he analyzed a variety of metric systems: topological spaces consisting of a well-defined set of points and a rule for measuring the "distance" between any two points.

Taxicab geometry is a metric system in which the points of the space correspond to the intersections of the horizontal and vertical lines of square-celled graph paper, or to the intersections of the streets in our idealized city. If two points, A and B, are at intersections on the same street, the distance between them is measured, as it is in Euclidean geometry, by counting the number of unit blocks from one to the other. If A and B are not on the same street, however, then instead of applying the Pythagorean theorem to calculate the distance between them we count the number of blocks a taxicab must travel as it goes from A to B (or vice versa) along a shortest-possible route. The structure of taxicab geometry can be formalized with definitions and axioms in a variety of ways, but here I shall dispense with such technicalities and simply describe it in intuitive terms.

In Euclidean geometry the minimum distance between two points (as the crow flies) defines a unique straight line. In taxicab geometry there may be many paths, all equally minimal, that join two points. In what follows "path" will be used to mean any taxicab route that covers the distance between two points with the minimum mileage.

If two points are not on the same street, how many distinct paths connect them? Pascal's famous number triangle comes to our aid on this question. Consider the points A and B at opposite corners of a 2-by-3 rectangle of blocks, as shown in Figure 63. The colored lines at the right in the illustration show how the rectangle can be drawn on Pascal's triangle to solve the problem. The lowest corner of the rectangle marks the answer: There are 10 distinct paths between A and B. Note that Pascal's triangle is left–right symmetrical, and so it does not matter in the least if the rectangle is drawn so that it leans the other way. The same answer is obtained. (Remember that in Pascal's triangle each number is the sum of the two numbers above it. For more on Pascal's triangle see Chapter 15 of my *Mathematical Carnival*.)

Figure 63

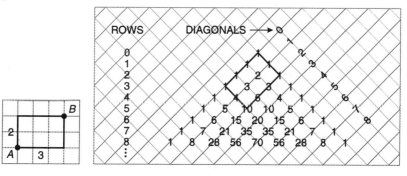

Readers familiar with combinatorics will recall how Pascal's triangle serves to instantly show how many ways a set of n items can be selected from a larger set of r items. The answer is the number at the intersection of the nth diagonal and the rth row of the triangle. In the case of the taxicab problem 10 is the number of ways two items can be selected from five. The two corresponds to one side of our rectangle of blocks and the five to the sum of its two sides. Ten is also the number of minimal routes a taxicab can follow from one corner of a 3-by-2 rectangle to the diagonally opposite corner.

It is not necessary to draw Pascal's triangle to determine the number of paths between two points in taxicab geometry. We can also use the familiar formula for calculating the number N of ways to select n objects from r objects: $N = r!/n!(r - n)!]$. For example, in our taxicab problem $r!$ equals $1 \times 2 \times 3 \times 4 \times 5$, or 120, $n!$ equals 1×2, or 2, and $(r - n)!$ equals $1 \times 2 \times 3$, or 6, so that the formula reduces to $N = 120/12$, or 10.

The fact that the rectangle can be tipped in either direction on Pascal's triangle is a pictorial way of saying that the number of ways of selecting n items from a larger set of r items is the same as the number of ways of selecting $r - n$ items from a set of r items. This fact becomes intuitively obvious if you consider that each time a unique set of n items is selected from r items, a unique set of $r - n$ items remains. In the taxicab model this means that if a Euclidean rectangle is drawn on the lattice, the number of distinct taxicab paths between any two diagonally opposite corners is the same as the number of paths joining the other two corners.

Since the "straight lines" (the shortest paths) of taxicab geometry may be crooked from the Euclidean point of view, the concept of an "angle" becomes either meaningless or radically different in this system. It is nonetheless possible to define close analogues of Euclidean polygons, including a two-sided "biangle" that is a stranger to Euclid's geometry. Some examples of biangles are shown in Figure 64. It should be obvious that although different biangles can share the same pair of "corner" points, the two "sides" of any biangle must be equal because they join the same two points.

A taxicab scalene triangle with corners A, B, and C and sides of 14, 8, and 6 is shown at the left in Figure 65. The sides of taxi polygons must of course be taxi paths, and the paths that make up a polygon of specified dimensions may vary in shape but not in length. Observe how the triangle in the illustration violates the Euclidean theorem that the sum of any two sides of a triangle must be greater than the third side. In this case the sum of two sides equals the third: 6 + 8 equals 14.

Figure 64

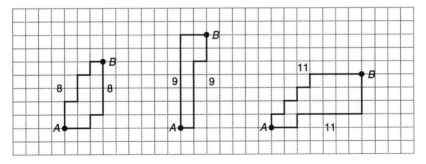

Figure 65

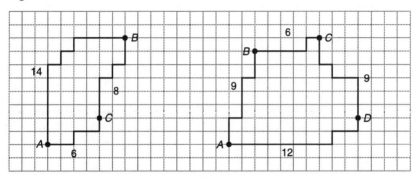

A taxi quadrilateral of sides 9, 6, 9, and 12 is shown at the right in the illustration.

Three taxi squares, all of side 6, are shown in Figure 66. Only the square at the left obeys the Euclidean theorem that the diagonals of a square are equal. As these figures demonstrate, taxicab squares can have innumerable Euclidean shapes.

It is easy to define a circle in taxicab geometry, and the result is quite unexpected. As in Euclidean geometry a circle is defined as the locus of all the points that are the same distance from a given point. Suppose the distance is 2. The resulting circle consists of the eight points shown at the left in Figure 67—a neat way to square the circle! Note that only one radius goes from the center point O to points A, B,

Figure 66

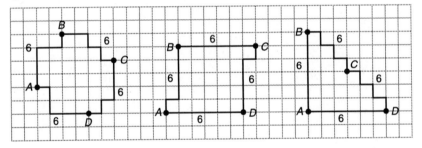

Figure 67

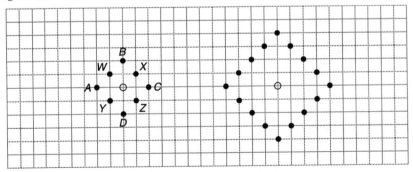

C, and *D*, and there are two radiuses going to each of the other four points. It is not hard to show that any taxicab circle of radius *r* consists of 4*r* points and has a circumference of 8*r*. If we adopt the Euclidean definition of pi as the ratio of the circumference of any circle to its diameter, then taxicab pi is exactly 4.

Observe that our eight-point taxicab circle includes a variety of sets of vertexes for taxi polygons of 2, 3, 4, 5, 6, 7, and 8 sides. For example, there is the biangle *D,X*, the equilateral triangle *B,C,D*, the square *A,B,C,D*, the regular pentagon *A,W,X,Z,Y*, the regular hexagon *A,W,B,X,Z,Y*, and the regular heptagon *A,W,X,C,Z,D,Y*. And the eight points of the circle lie at the corners of a set of regular octagons.

Another Euclidean theorem that taxicab geometry violates is the one stating that two circles can intersect at no more than two points. As shown in Figure 68, two taxicab circles may intersect at any finite number of points. The larger the circles are, the more points at which they can intersect. A little experimentation turns up excellent taxicab analogues of the other three conic-section curves. Figure 69 shows four 12-point taxicab ellipses. As in Euclidean geometry, a taxicab ellipse is the locus of points whose distances from two fixed points *A* and *B* have the same sum. The points, called foci, are marked here with colored circles, and in all the examples shown in the illustration the constant sum is 6.

The fourth curve is actually a degenerate ellipse corresponding to the straight line that results when the constant sum that defines a Euclidean ellipse equals the distance between its foci. If this equality holds in taxicab geometry, then when *A* and *B* are on the same street, the result is a straight line of points. Otherwise the ellipse consists of all the points within the Euclidean rectangle of lattice lines that has *A* and *B* at diagonally opposite corners. For example, suppose that *A* and *B* are opposite corners of a square with lattice sides of length 4.

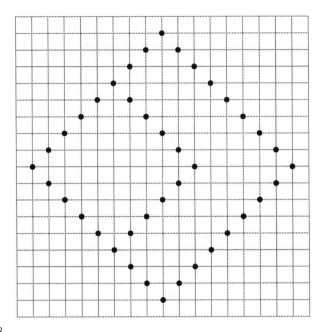

Figure 68

Figure 69

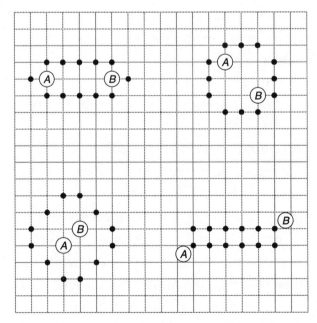

In this case the taxi distance between A and B is 8, and for each of the square's 25 points the sum of their distances from A and B is 8. These 25 points will be the degenerate ellipse of constant sum 8 whose foci are A and B. If the constant sum is greater than the taxi distance between A and B, then as in Euclidean geometry the taxi ellipse becomes more circular as the foci move closer together. When A and B coincide, then once again as in Euclidean plane geometry the ellipse becomes a circle.

A Euclidean parabola is the locus of all points whose distance from a focus A is equal to its shortest distance from a fixed straight line: the directrix. If a taxicab directrix is defined as the set of points along a Euclidean straight line, then taxicab parabolas can also be constructed. Two are shown at the left in Figure 70. Try drawing the parabola for the directrix and the focus shown at the right.

Taxicab hyperbolas are more complex. A Euclidean hyperbola is the locus of all points for which the difference between the distances from a pair of foci A and B is constant. The appearance of a taxicab hyperbola varies considerably as the ratio of its basic parameters varies. In the figure at the left in Figure 71, the foci A and B are placed to show the limiting case, a degenerate hyperbola of just one branch, where the constant difference is 0. The figure at the right in the illustration shows two infinitely long branches of a taxicab hyperbola with a constant difference of 4.

Taxicab geometry springs another surprise in Figure 72. In this hyperbola the constant difference is 2. Here the two branches are two infinite sets of points, one in the sector of the plane at the upper left and one in the sector at the lower right, and each with a "tail" of infinite length. As shown at the bottom in the illustration, the results are similar when the constant is 8, except that the infinite sets of points are in the sectors of the plane at the upper right and the lower left, and there are no tails.

Figure 70

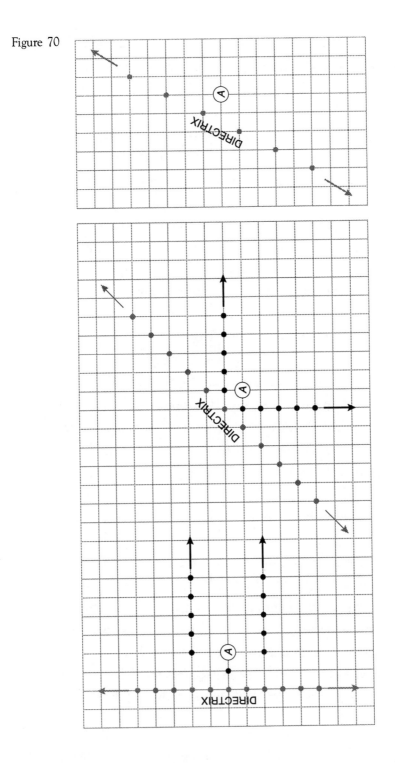

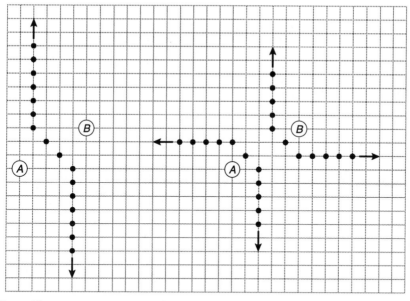

Figure 71

With the foci placed as shown in all of these examples the constant difference cannot be an odd number because the resulting figure would include points away from intersections, the only allowable points in taxicab space. As an exercise place A and B at diagonally opposite corners of a 3-by-6 Euclidean rectangle of lattice sides and draw the hyperbola for which the constant difference is 1. The result is two "parallel" branches, each resembling the degenerate hyperbola with a constant difference of 0. A not-so-easy problem is to define the exact conditions under which taxicab hyperbolas of the five general types are created.

Only one book has been published on taxicab geometry: it is *Taxicab Geometry*, a paperback by Eugene F. Krause, a mathematician at the University of Michigan. (This work along with a few of the dozen or so papers on the topic that have appeared in British mathematical journals over the past two decades are listed in the bibliography at the end of this chapter) Krause's book is recommended particularly to

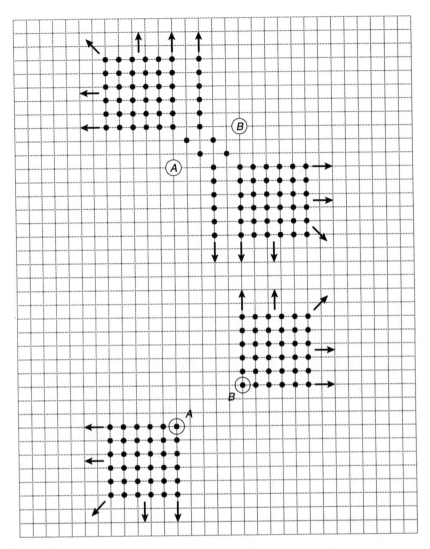

Figure 72

students who want to learn how taxicab geometry can be elegantly generalized to the entire Cartesian plane, where all points are represented by ordered pairs of real numbers from the two coordinate axes. The rule of measuring distance by the shortest path along line segments that parallel the axes must of course be preserved, so in this continuous form of taxicab geometry an infinite number of distinct paths, all of the same minimum length, connect any two points that are not on the same street.

Krause shows how continuous taxicab geometry satisfies all but one of the postulates of Euclidean geometry. Instead of violating the notorious parallel postulate, as elliptic and hyperbolic geometries do, taxicab geometry violates the side-angle-side postulate, which states that two triangles are congruent if and only if two sides and the included angle of one are congruent to two sides and the included angle of the other.

Midway between the discrete taxicab geometry I have described (which is confined to what is often called the lattice of integers) and the continuous version is another taxicab geometry in which the points of the associated space are defined by ordered pairs of rational numbers. Even on the lattice of integers, however, taxicab geometry provides a fertile field for investigation by recreational mathematicians and should present a splendid and enriching challenge for high school students. I have barely scratched the surface here, leaving many fundamental questions unanswered. How should parallel lines be defined? What is the best analogue of a perpendicular bisector? Are there useful ways to define area?

Moreover, taxicab geometry extends readily to integer lattices of three dimensions and higher. The exploration of taxicab geometries on other kinds of lattices, such as triangular or hexagonal ones that are either finite or infinite, is still a wide-open field. Indeed, the lattices need not be confined to a plane. They can be defined on the surface of cylinders, spheres, toruses, Möbius bands, Klein bottles—anything you like! Just make sure your cabbies stick to the streets and always take you by the shortest path to where you want to go.

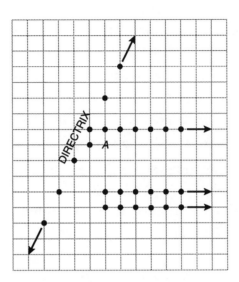

Figure 73

Answers

The solution to the problem of constructing a taxicab parabola with a given focus A and a given directrix is shown in Figure 73.

ADDENDUM

Kenneth W. Abbott, a New York computer consultant, sent an amusing generalization of discrete "taxicab" geometry. As in taxicab geometry, the points of the non-Euclidean space are the intersections of lattice lines on a square grid. In Abbott's generalization the "distance" between any two points is an integer defined as being equal to $\sqrt[n]{x^n + y^n}$, where x is measured horizontally, y is measured vertically and n is any positive integer.

When n is 1, we have the simple taxicab geometry explained in this chapter. All "circles" are sets of points equidistant from the circle's center. They have the forms shown at the left in Figure 74, where the radii are 1, 2, 3, 4, and 5.

When n is 2, circles of the same radii take the forms shown in the middle of the illustration. Note that the first four circles consist of just four points lying on the two axes that pass through the common center of the five circles. We shall call such circles "trivial." When n is 1, only the circle of radius 1 is trivial. All other circles are not. When n is 2, the fifth circle is nontrivial. In this geometry there are an infinite number of both kinds of circles. Pi is $2\sqrt{2}$ for all trivial circles, but it has different values for the nontrivial ones. For the fifth circle, which has a radius of 5, pi is $(4\sqrt{10} + 2\sqrt{2}) / 5$.

When n is 3, the first five circles [*at right in Figure 74*] are all trivial. In this geometry pi is $2^{(n+1)/n}$ for all trivial circles.

We now state a remarkable theorem. Any generalized taxicab geometry with n greater than 2 can contain only trivial circles. This assertion as Abbott pointed out, is easily seen to be equivalent to Fermat's last theorem!

Figure 74

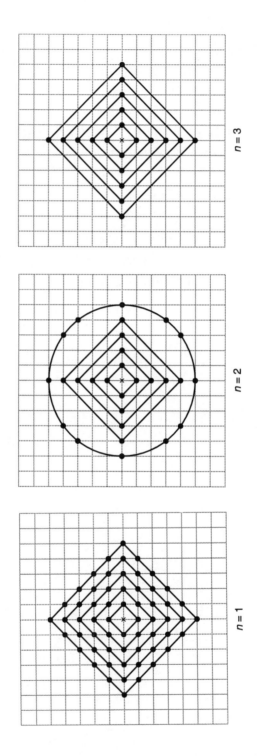

References

SQUARE CIRCLES. Francis Sheid in *The Mathematics Teacher*, Vol. 54, No. 5, pages 307–312; May 1961.

SQUARE CIRCLES. Michael Brandley, in *The Pentagon*, pages 8–15; Fall 1970.

TAXICAB GEOMETRY—A NON-EUCLIDEAN GEOMETRY OF LATTICE POINTS. Donald R. Byrkit in *The Mathematics Teacher*, Vol. 64, No. 5, pages 418–422; May 1971.

TAXICAB GEOMETRY. Eugene F. Krause. Addison-Wesley Publishing Company, 1975. Dover reprint 1986.

TAXICAB GEOMETRY. Barbara E. Reynolds in *Pi Mu Epsilon Journal*, Vol. 7, No. 2, pages 77–88; Spring 1980.

PYRAMIDAL SECTIONS IN TAXICAB GEOMETRY. Richard Laatsch in *Mathematics Magazine*, Vol. 55, pages 205–212; September 1982.

LINES AND PARABOLAS IN TAXICAB GEOMETRY. Joseph M. Moser and Fred Kramer in *Pi Mu Epsilon Journal*, Vol. 7, pages 441–448; Fall 1982.

TAXICAB GEOMETRY: ANOTHER LOOK AT CONIC SECTIONS. David Iny in *Pi Mu Epsilon Journal*, Vol. 7, pages 645–647; Spring 1984.

THE TAXICAB GROUP. Doris Schattschneider in *American Mathematical Monthly*, Vol. 91, pages 423–428; August–September 1984.

TAXICAB TRIGONOMETRY. Ruth Brisbin and Paul Artola in *Pi Mu Epsilon Journal*, Vol. 8, pages 89–95; Spring 1985.

A FOURTH DIMENSIONAL LOOK INTO TAXICAB GEOMETRY. Lori J. Mertens in *Journal of Undergraduate Mathematics*, Vol. 19, pages 29–33; March 1987.

TAXICAB GEOMETRY—A NEW SLANT. Katye O. Sowell in *Mathematics Magazine*, Vol. 62, pages 238–248; October 1989.

KARL MENGER AND TAXICAB GEOMETRY. Louise Golland in *Mathematics Magazine*, Vol. 63, pages 326–327; October 1990.

II

The
Power
of the
Pigeonhole

If m male pigeons have sex with n
female pigeons and $m > n$, then at
least two male pigeons must have
sex with the same female pigeon.

—Anonymous

Can you prove that a large number of people in the U.S. have
exactly the same number of hairs on their head? And what
does this question have in common with the following problem? In a bureau drawer there are 60 socks, all identical except for their
color: 10 pairs are red, 10 are blue, and 10 are green. The socks are all
mixed up in the drawer, and the room the bureau is in is totally dark.
What is the smallest number of socks you must remove to be sure you
have one matching pair?

Consider two less simplistic examples. Can you prove that when a common fraction a/b is expressed in decimal form, the resulting number will be either a terminating decimal or one that repeats with a period no longer than b? Can you show that if five points are placed anywhere on an equilateral triangle of side 1, at least two points will be no farther apart than .5? (Hint: Divide the triangle into four smaller equilateral triangles of side .5.)

The quality these problems and thousands of others in both serious and recreational mathematics have in common is that they can all be solved by invoking an old and powerful principle. It is the pigeonhole principle, which some mathematicians prefer to call the Dirichlet drawer principle after the 19th-century German mathematician Peter Gustav Lejeune Dirichlet. The pigeonhole principle is the topic of this chapter—a chapter written not by me but by Ross Honsberger, a mathematician at the University of Waterloo. He is the author of *Ingenuity in Mathematics, Mathematical Gems, Mathematical Gems II, Mathematical Gems III*, and has edited the anthologies *Mathematical Morsels, More Mathematical Morsels, Mathematical Plums*, and most recently, *Episodes in Nineteenth and Twentieth Century Euclidean Geometry*. All eight are excellent sources of unusual problems with a strong recreational flavor. Everything that follows (up to my concluding comments) was written by Honsberger, who calls his discussion of the pigeonhole principle "Can anything this simple be useful?"

Consider the statement "If two integers add up to more than 100, at least one of them is greater than 50." It is far from obvious that the "overflow" principle behind this simple assertion is not trivial. In its simplest form the principle can be stated as follows: If $n + 1$ (or more) objects are to be distributed among n boxes, some box must get at least two of the objects. More generally, if $kn + 1$ (or more) objects are distributed among n boxes, some box must get at least $k + 1$ of the objects.

Even in its most general form this pigeonhole principle states for a set of data only the obvious fact that it is not possible for every value to be below average or for every value to be above average. Nevertheless, the principle is a mathematical concept of major importance and remarkable versatility. Here we shall take up seven of its prettiest elementary applications. Let us begin with a simple geometric example.

1. *The Faces of a Polyhedron.* Try counting the edges around the faces of a polyhedron. You will find that there are two faces bounded by the same number of edges. To prove that this is always the case it is only necessary to imagine what happens when the faces are distributed among a set of boxes numbered 3, 4, . . . n, so that a face with r edges is put in the box numbered r. Since edges separate faces, a face with the maximum number of edges n is itself bordered by n faces, implying that the polyhedron must have a total of at least $n + 1$ faces. On the pigeonhole principle, then, some box must contain at least two of the faces, and the proof is complete. In fact, it is a simple exercise to show that there are always at least two different pairs of faces with the same number of edges.

2. *Ten Positive Integers Smaller than 100.* Here is an application of the pigeonhole principle that will baffle your friends. No matter how a set S of 10 positive integers smaller than 100 is chosen there will always be two completely different selections from S that have the same sum. For example, in the set 3, 9, 14, 21, 26, 35, 42, 59, 63, 76 there are the selections 14, 63, and 35, 42, both of which add up to 77; similarly, the selection 3, 9, 14 adds up to 26, a number that is a member of the set.

To see why this is always the case observe that no 10-element subset of S can have a sum greater than the 10 largest numbers from 1 to 100: 90, 91, . . . 99. These numbers add up to 945, and so the subsets of S can be sorted according to their sum into boxes numbered 1, 2, . . . 945. Since each member of S either belongs to a particular subset or does not belong to it, the number of subsets to be classified (not counting the empty set, which has no elements) is $2^{10} - 1$, or 1,023. By the pigeonhole principle, then, some box must contain (at least) two different subsets A and B. Discarding any numbers that are in both A and B creates two disjoint subsets, A' and B', with equal sums. Indeed, because there are 78 more subsets than there are boxes, every set S must actually yield dozens of different pairs of subsets with equal sums.

3. *The Pills.* For this next application of the pigeonhole principle we are indebted to Kenneth R. Rebman, a mathematician at California State University at Hayward. A physician testing a new medication instructs a test patient to take 48 pills over a 30-day period. The patient is at liberty to distribute the pills however he likes over this period as long as he takes at least one pill a day and finishes all 48 pills by the end of the 30 days. No matter how the patient decides to arrange things, however, there will be some stretch of consecutive days in which the total number of pills taken is 11. In fact, for every value of k from 1 to 30 except 16, 17, and 18 it is always possible to find a period of consecutive days in which a total of k pills were taken.

To prove that a particular value of k is an exception to the rule it is necessary only to find a distribution of the pills for which there is no stretch of days when k pills are taken. Thus the cases $k = 16$, $k = 17$, and $k = 18$ are eliminated at a stroke by the following distribution in which one pill is taken each day except for the 16th, when 19 pills are taken:

$$\underbrace{11\ldots1}_{15}\ 19\ \underbrace{11\ldots1}_{14}$$

Now consider the case $k = 11$. If p_i denotes the total number of pills that have been taken up to the end of the ith day, then p_{30} equals 48 and the positive number $p_1, p_2, \ldots p_{30}$ form a strictly increasing sequence $0 < p_1 < p_2 < \ldots < p_{30} = 48$. (The sign "<" is read "less than," and a strictly increasing sequence is one where each element is larger than its predecessor.) Adding 11 to each of the numbers in this sequence creates a new strictly increasing sequence: $11 < p_1 + 11 < p_2 + 11 < \ldots < p_{30} + 11 = 59$.

There are 30 numbers p_i in the first sequence and 30 numbers $p_i + 11$ in the second, and all 60 of these positive numbers are less than or equal to 59. Therefore by the pigeonhole principle at least two of them must be equal. No two p_is are the same, however, and as a result no two $p_i + 11$s are the same. Hence some p_i must be equal to some $p_j + 11$, that is $p_i = p_j + 11$ for some values of i and j. Hence p_i minus p_j equals 11, which implies that precisely 11 pills were taken on the consecutive days $j + 1, j + 2, \ldots i$.

This argument holds for any value of k up to and including 11, establishing the property for the entire block of values of k from 1 through 11. It is somewhat more complicated to dispose of the remaining cases, but the pigeonhole principle is the critical tool throughout. Consider next the cases $k = 31$ through $k = 47$. Although these values of k certainly admit of solutions in many instances, the following family of distributions shows that no one of them guarantees a solution. When n is between 1 and 17, the value $k = 30 + n$ is eliminated by the following sequence:

$$(19 - n) \underbrace{11 \ldots 1}_{n+11} (n+1) \underbrace{11 \ldots 1}_{17-n}$$

For example, when n equals 7, the following distribution eliminates the case $k = 37$.

$$12 \underbrace{11 \ldots 1}_{18} 8 \underbrace{11 \ldots 1}_{10}$$

4. *101 Numbers.* Suppose some set of 101 numbers $a_1, a_2, \ldots a_{101}$ is chosen from the numbers 1, 2, . . . 200. Surprisingly, it turns out to

be impossible to choose such a set without taking two numbers for which one divides the other evenly, that is, with no remainder. Proving that this assertion is true provides an opportunity to make use of a rather neat way of expressing integers.

Given a positive integer n, it is possible to factor out of it as many 2s as it contains in order to reduce it to the form $n = 2^r q$, where q is an odd integer (possibly as small as 1). If each of the selected numbers a_i is expressed in this form, a set of 101 values of q is obtained, each value belonging to the set of 100 odd numbers 1, 3, 5, . . . 199. On the pigeonhole principle it can be concluded that two of these values of q must be the same. Therefore for some integers i and j, a_i equals $2^{r_i}q$ and a_j equals $2^{r_j}q$. Of these two numbers the one with the smaller power of 2 clearly divides the other.

Similarly, it is not difficult to apply the pigeonhole principle to show that any set S consisting of 102 numbers from the set 1, 2,. . . 200 must have two distinct numbers that add up to a third number in S. (Here it is not necessary to employ the form $2^r q$.) I shall turn next to two spectacular applications of the pigeonhole principle in geometric settings.

5. *Six Hundred and Fifty Points in a Circle.* Consider a circle C with a radius of 16 and an annulus, or ring A, with an outer radius of 3 and an inner radius of 2. Is it not remarkable that wherever one might sprinkle a set S of 650 points inside C the annulus A can always be placed on the figure so that it covers at least 10 of the points? To demonstrate the truth of this assertion one could place 650 copies of the ring A on the region enclosed by the circle C so that each point of S was the center of one of the rings, as is suggested in Figure 75. For points of S near the circumference of C the corresponding annuli will extend beyond the circle. On the other hand, a circle concentric with C that has radius 19 (equal to the radius of C plus the outer radius of A) will certainly enclose all the copies of A. Call

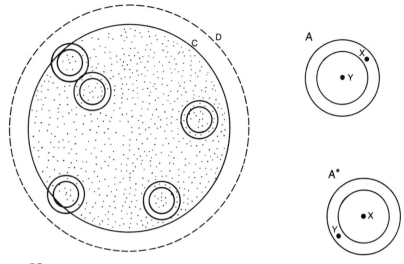

Figure 75

this new circle D. Note that the area of D is $\pi 19^2$, or 361π, and since the area of A is $\pi 3^2 - \pi 2^2$, or 5π, then 650 copies of A have a total area of $3{,}250\pi$.

It is at this point that a "continuous" version of the pigeonhole principle can be applied. Each copy of A covers certain parts of the circle D when it is placed on that figure. Suppose when all 650 copies have been put in place, there is no part of D that lies under more than nine different copies of A. In that case the total area of the copies could not exceed nine times the area of D. This, however, cannot be the case, because $9(361\pi)$ is only $3{,}249\pi$, whereas the annuli have a total area of $3{,}250\pi$. The pigeonhole principle, then, implies that some point X of D must be covered by at least 10 copies of A.

Now suppose Y is a point of S that is at the center of one of these 10 copies of A. Then the distance from X to Y must be larger than the inner radius of A and smaller than the outer radius, and as is shown at the right in Figure 75 another copy of A centered at X would cover Y. Call this copy A^*. Since there are at least nine other centers like Y, A* must cover at least 10 points of S, and the assertion is proved. (This problem was proposed by Viktors Linis of the University of Ottawa in *Crux Mathematicorum*, Vol. 5, page 271; November 1979.)

6. *The Marching Band*. The next example concerns a marching band whose members are lined up in a rectangular array of m rows and n columns. Viewing the band from the left side, the bandmaster notices that some of the shorter members are hidden in the array. He rectifies this aesthetic flaw by arranging the musicians in each row in nondecreasing order of height from left to right, so that each one is of height greater than or equal to that of the person to his left (from the viewpoint of the bandmaster). When the bandmaster goes around to the front, however, he finds that once again some of the shorter members are concealed. He proceeds to shuffle the musicians within their columns so that they are arranged in nondecreasing order of height from front to back. At this point he hesitates to go back to the left side to see what this adjustment has done to his carefully arranged rows. When he does go, however, he is pleasantly surprised to find that the rows are still arranged in nondecreasing order of height from left to right! Shuffling an array within its columns in this manner does not undo the nondecreasing order in its rows.

This startling fact can be proved indirectly, by assuming that it is false and arriving at a contradiction. In other words, we shall assume that after the columns have been arranged there is a row in which a taller musician a is placed ahead, or to the left, of a shorter one b. Call the column the taller musician a is in i and the column the shorter musician b is in j, as is shown in Figure 76. Since the columns have just been arranged, it can be assumed that every musician in the segment P from a back in column i is at least as tall as a and that every musician in the segment Q from b forward in column j is no taller

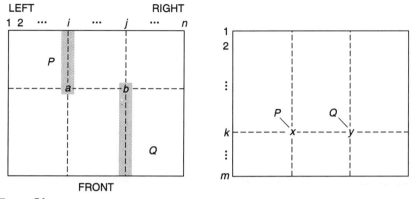

Figure 76

than b. Moreover, since a is taller than b, it follows that every member of P is taller than every member of Q.

Now consider the halfway point at which the rows have been arranged but not the columns. To get back to this point it is necessary to return the musicians in segment P to their former positions throughout column i and to return those in segment Q to their former positions throughout column j. In other words, the members of P and Q will be distributed over the rows 1, 2, ... m as if the m rows were boxes. Segments P and Q, however, have a total length of $m + 1$, that is, there is a total of $m + 1$ musicians in the two segments. On the pigeonhole principle two of the musicians must end up in the same row. They could not both have come from the same segment, and so in some row there must be a member x from P in column i ahead, or to the left, of a member y from Q in column j, as is shown at the right in Figure 76. Since x is taller than y, this arrangement violates the already established nondecreasing order of the rows, and the conclusion follows by contradiction.

7. *Subsequences in a Permutation.* This final example establishes an engaging property of every arrangement of numbers from the sequence

1, 2, . . . $n^2 + 1$ into a row. When each arrangement is scanned from left to right, it must contain either an increasing subsequence of length (at least) $n + 1$ or a decreasing subsequence of length (at least) $n + 1$. For example, when n equals 3, the arrangement 6, 5, 9, 3, 7, 1, 2, 8, 4, 10 includes the decreasing subsequence 6, 5, 3, 1. (As this example demonstrates, a subsequence need not consist of consecutive elements of the arrangement.)

The assertion that every arrangement includes a subsequence of this kind can be demonstrated easily by specifying that for each number i in the row, x stands for the length of the longest increasing subsequence that begins with i, and y stands for the length of the longest decreasing subsequence that begins with i.

In this way $n^2 + 1$ pairs of "coordinates" (x, y) are obtained for the row of numbers, and if any value of x or y is as great as $n + 1$, the assertion is valid. On the other hand, if every value of x and y is less than or equal to n, there are only n^2 possible different pairs (x, y). In this case the pigeonhole principle implies some pair (x, y) would have to be the coordinates of at least two numbers i and j in the row. But i is not equal to j, and if i is less than j, then the x coordinate of i would be greater than that of j, and if i is greater than j, then the y coordinate of i would be greater than that of j. In either case a contradiction has been reached and so the assertion is proved.

Let us close with three exercises the reader may enjoy.

1. A lattice point is a point in a coordinate plane for which both coordinates are integers. Prove that no matter what five lattice points might be chosen in the plane at least one of the segments that joins two of the chosen points must pass through some lattice point in the plane.

2. Six circles (including their circumferences) are arranged in the plane so that no one of them contains the center of another. Prove that they cannot have a point in common.

3. Prove that in any row of $mn + 1$ distinct real numbers there must be either an increasing subsequence of length (at least) $m + 1$ or a decreasing subsequence of length (at least) $n + 1$.

The counterintuitive result Honsberger describes for the marching band can be demonstrated dramatically with a deck of playing cards. Shuffle and deal the cards face up in any rectangular array, say four rows of six cards each as is shown in Figure 77, top, and then rearrange the cards in each row so that from left to right the six values never decrease, as is shown in the middle of the illustration. (For example, an acceptable arrangement is 6, 7, 10, 10, J, K.) Now rearrange each column so that from top to bottom the four values never decrease, as is

Figure 77

Q ♥	5 ♥	K ♦	9 ♣	4 ♥	7 ♣
3 ♣	5 ♦	A ♦	4 ♦	6 ♦	A ♠
5 ♠	3 ♠	3 ♦	2 ♣	9 ♠	8 ♣
10 ♦	10 ♣	7 ♦	J ♠	6 ♥	K ♠

4 ♥	5 ♥	7 ♣	9 ♣	Q ♥	K ♦
A ♦	A ♠	3 ♣	4 ♦	5 ♦	6 ♦
2 ♣	3 ♠	3 ♦	5 ♠	8 ♣	9 ♠
6 ♥	7 ♦	10 ♦	10 ♣	J ♠	K ♠

A ♦	A ♠	3 ♣	4 ♦	5 ♦	6 ♦
2 ♣	3 ♠	3 ♦	5 ♠	8 ♣	9 ♠
4 ♥	5 ♥	7 ♣	9 ♣	J ♠	K ♦
6 ♥	7 ♦	10 ♦	10 ♣	Q ♥	K ♠

shown at the bottom. Permuting the cards in a column will of course alter the cards in the rows. Nevertheless, after ordering the columns you will find that the rows remain ordered!

A card trick based on this surprising finding appeared in the magic periodical *The Pallbearers Review* (page 513, April 1972). Deal five poker hands and then rearrange each hand so that its five cards are in increasing order from back to front. Assemble the hands any way you like and then deal five new hands face down in the conventional manner. The hands will be entirely different and will not be ordered. Explain that you are trying to teach the cards to order themselves. Pick up each hand, order its cards once more, and turn the hand face down. Assemble the hands by placing the fifth hand (the dealer's) on top of the fourth hand, those two hands on top of the third, and so on. Deal five more hands face down in the usual way. The cards will have learned their lesson: although once again the hands will all be altered, each hand will be ordered!

This result is part of the theory of Young Tableux, a class of number arrays named for the Reverend Alfred Young, the British clergyman who proposed and analyzed them in a paper published in 1900. The arrays have been shown to have important applications in quantum mechanics.

In the early 1960s the marching-band problem appeared in various guises in several mathematics journals. David Gale and Richard M. Karp wrote a monograph on the subject titled "The Nonmessing-up Theorem," published in 1971 by the operations research center of the engineering school of the University of California at Berkeley.

Donald Knuth, in the third volume of his classic *Art of Computer Programming*, discusses the theorem in connection with a method of sorting called "shellsort."

Answers

1. To show that one of the line segments connecting five lattice points must pass through some lattice point in the coordinate plane, note that there are four "parity" classes for the coordinates of a lattice point: odd, odd; odd, even; even, odd; and even, even. On the pigeonhole principle some two of five lattice points, say (x_1, y_1) and (x_2, y_2), must belong to the same class. This implies that $x_1 + x_2$ and $y_1 + y_2$ are both even numbers, making the midpoint of the segment joining the points, namely $[(x_1 + x_2)/2, (y_1 + y_2)/2]$, a lattice point.

2. To prove that six circles arranged in the plane so that none of them contains the center of another cannot have a point in common, assume the converse is true, namely, that there is a point O common to six such circles. Now suppose O is joined to each of the six centers. No two centers can be colinear with O because no circle contains the center of another circle and all the circles contain O. Therefore the six lines all fan out from O. Let OA and OB be consecutive segments in the fan. Since O belongs to each circle, the segments OA and OB are not larger than the radii of the circles in which they lie. But since neither circle contains the other center, AB must be larger than either of these radii. Thus AB is longer than the other two sides of triangle AOB, which implies that angle AOB opposite AB is larger than either of the other angles in the triangle. Hence angle AOB must exceed 60 degrees. If this is so, however, there is not room in the 360-degree sweep around O for six angles such as AOB, which establishes the conclusion by contradiction.

3. To prove that in any row of $mn + 1$ distinct real numbers there is either an increasing subsequence of length $m + 1$ or a decreasing subsequence of length $n + 1$ let "coordinates" (x, y) be assigned as in example 7. The conclusion holds either if x is greater

than m or if y is greater than n. Now, when x is less than or equal to m and y is less than or equal to n, there are only mn different pairs (x, y). On the pigeonhole principle two of the pairs assigned to the $mn + 1$ numbers in the row must be the same, and as was shown, a contradiction follows.

ADDENDUM

One of the simplest examples of how quickly the pigeonhole principle solves a geometrical problem—Honsberger did not mention it because it is so well known—is to prove that if five points are in or on a square of side 1, then at least one pair of points will be no farther apart than half the square root of 2.

Divide the square into four squares, each of side 1/2. By the pigeonhole principle, one of the four squares must contain two of the five points. Because the small square's diagonal is half the square root of 2, the two points must be that distance apart or less.

References

THE PIGEONHOLE PRINCIPLE: "THREE INTO TWO WON'T GO." Richard Walker in *The Mathematical Gazette*. Vol. 61, No. 415, pages 25-31; March 1977.

EXISTENCE OUT OF CHAOS. Sherman K. Stein in *Mathematical Plums: The Dolciani Mathematical Expositions*. No. 4, edited by Ross Honsberger. The Mathematical Association of America, 1979.

THE PIGEONHOLE PRINCIPLE. Kenneth R. Rebman in *The Two-Year-College Mathematics Journal*, mock issue, pages 4-12; January 1979.

PIGEONS IN EVERY PIGEONHOLE. Alexander Soifer and Edward Lozansky in *Quantum*, pages 25-26, 32; January 1990.

NO VACANCY. Dominic Olivastro in *The Sciences*, pages 53-55; September/October 1990.

APPLICATIONS OF THE PIGEON-HOLE PRINCIPLE. Kiril Bankov in *The Mathematical Gazette*, Vol. 79, pages 286-292; May 1995.

12

Strong Laws of Small Primes

Let us now praise prime numbers
With our fathers that begat us:
The power, the peculiar glory
    of prime numbers
Is that nothing begat them,
No ancestors, no factors.
Adams among the multiplied
    generations.

          –Helen Spalding

"The Strong Law of Small Numbers" is the provocative title of an unpublished paper by Richard Kenneth Guy, a mathematician at the University of Calgary. For many years Guy has edited the "Research Problems" department of *The American Mathematical Monthly*. He is the author of numerous technical papers and is coauthor with John Horton Conway and Elwyn R. Berlekamp of *Winning Ways*, a two-volume work about new mathematical recreations. The material that follows is taken almost entirely from Guy's paper.

"We think of mathematics as an exact science," Guy begins, "but in the field of discovery this is not at all the right picture. Two of the most important elements in mathematical research are asking the right questions and recognizing patterns."

Unfortunately there is no procedure for generating good questions and no way of knowing whether an observed pattern will lead to a significant new theorem or whether the pattern is just a lucky coincidence. In these respects the research mathematician is in a position strangely like that of the scientist. Both ask questions, do experiments and observe patterns. Will an observed pattern be repeated when new observations are made, with new parameters, leading to the discovery of a general law, or will counterexamples turn up that contradict a hypothesis? It is true that mathematicians can do something scientists cannot: they can prove theorems within a formal system. Until a proof is found, however, a mathematician relies on fallible empirical induction in much the same way a scientist does. This is particularly true in combinatorial problems that involve infinite sequences of numbers.

In examining cases involving small numbers a striking pattern may be encountered that strongly implies a general theorem. It is this implication Guy calls the strong law of small numbers. Sometimes the law works, sometimes it does not. If the pattern is no more than a set of coincidences, as it often is, a mathematician can waste an enormous amount of time trying to prove a false theorem. The law can also mislead in an opposite way. A few counterexamples may cause the mathematician to prematurely abandon a search for a theorem that is actually true but slightly more complicated than expected.

Today's computers are a big help because they often can quickly explore cases of higher numbers that will either explode a hypothesis or greatly increase the probability of its being true. In many combinatorial problems, however, the numbers grow at such a fantastic rate that the computer can examine only a few more cases than can be examined by hand, and the mathematician may be left with an extremely intractable problem.

One could fill many books with examples of how the strong law of small numbers has led to significant theorems, or has misled investigators into looking for theorems that are not there, or has deceived them by suggesting a theorem is not there when it is, or has suggested theorems that may be there but resist all efforts to prove them. In the ragbag of examples that follow we shall limit our attention to positive prime numbers.

Primes are the natural numbers larger than 1 with no factors except 1 and themselves: 2, 3, 5, 7, 11, 13, 17, 19, 23, 29, 31, . . . All are odd except 2, which has the reputation, Guy points out, of being the "oddest" of all the primes. No simple formula generates all the primes and only primes. As the second stanza of Helen Spalding's poem goes:

> None can foretell their coming.
> Among the ordinal numbers
> They do not reserve their seats,
>   arrive unexpected.
> Along the line of cardinals
> They rise like surprising pontiffs,
> Each absolute, inscrutable,
>   self-elected.

Euclid proved that the primes are infinitely many, but the higher they go the larger the gaps between them are. The same is true of the prime powers. Apart from 6 every natural number smaller than 10 is a power of a prime, and more than a third of all numbers smaller than 100 are prime powers. Yet it would be folly to conclude from these small primes that the density of prime powers has a lower bound. They thin out so rapidly as the numbers get larger that their density can be made as low as one likes.

In the beginning where chaos
Ends and zero resolves,
They crowd the foreground prodigal
    as forest,
But middle distance thins them,
Far distance to infinity
Yields them rarely as unreturning
    comets.

Primes offer rich examples of remarkable patterns that are entirely accidental and lead nowhere. Consider the following sequence of primes 3, 31, 331, 3331, 33331, 333331, 3333331, 33333331. One is tempted to think the pattern will continue, but it fails in the next case: 333333331 is composite (nonprime) with the prime factors of $17 \times 19,607,843$. Indeed, in all cases of patterns of this kind it is a safe bet that the pattern will not continue to yield primes. Wade Philpott and Joe Reitch, Jr., checked the 3333. . . . 1 pattern for runs of 9 through 14 threes and found all six numbers to be composite.

Several years ago Reo F. Fortune, an anthropologist at the University of Cambridge (who was once married to Margaret Mead), noted a curious pattern involving small primes. Starting with 2, take the product of a set of consecutive primes. Add 1. Find the next largest prime and from it subtract the product of the consecutive primes. Is the result always a prime? The chart in Figure 78 shows the procedure applied to the first eight cases and gives the eight "fortunate primes" that are generated.

Figure 78

$$2 + 1 = 3$$
$$(2 \times 3) + 1 = 7$$
$$(2 \times 3 \times 5) + 1 = 31$$
$$(2 \times 3 \times 5 \times 7) + 1 = 211$$
$$(2 \times 3 \times 5 \times 7 \times 11) + 1 = 2311$$
$$(2 \times 3 \times \ldots \times 13) + 1 = 30031$$
$$(2 \times 3 \times \ldots \times 17) + 1 = 510511$$
$$(2 \times 3 \times \ldots \times 19) + 1 = 9699691$$

$$5 - 2 = 3$$
$$11 - 6 = 5$$
$$37 - 30 = 7$$
$$223 - 210 = 13$$
$$2333 - 2310 = 23$$
$$30047 - 30030 = 17$$
$$510529 - 510510 = 19$$
$$9699713 - 9699690 = 23$$

Fortune conjectures that the result is always a prime. Most number theorists believe this is true, but no proof has been found and there seems to be little hope, Guy says, of finding one in the foreseeable future. Perhaps some reader of this column can "cook" (falsify) the conjecture by finding what one might call a "fortune cookie." Note that in the chart the first five numbers at the right side of the equation on the left are primes. Is this always the case? No, it fails for the next three numbers. Mark Templer in his article titled "On the Primality of $k! + 1$ and $2 \cdot 3 \cdot 5 \cdot \ldots \cdot p + 1$" has shown (*Mathematics of Computation*, Vol. 34, No. 149, pages 303–304; January 1980) that one more than the product of primes up to p is prime for the first five primes and for $p = 31$, $p = 379$, $p = 1019$ and $p = 1021$, and for no other p less than 1032. [In a letter received after this chapter appeared in *Scientific American*, R. E. Crandall extended the list of primes to 2657, but for no other primes less than 3000.]

Another strange hypothesis, not yet proved, is known as the Gilbreath conjecture after Norman L. Gilbreath, an American mathematician and amateur magician who proposed it in 1958. Write the sequence of primes in a row and under them list the differences between successive primes. Under that second row list the absolute values of the differences, and continue the procedure for as long as you like. Figure 79 shows a table of nine rows of differences for the first 24 primes. Note that each row begins with 1. Will every row begin with 1? Gilbreath guesses that it will. This has been verified by Ray B. Killgrove and Ken E. Ralston up to the 63,419th prime (*Mathematical Tables and Other Aids to Computation*, Vol. 13, No. 66, pages 121–122; April 1959).

"It does not seem likely," writes Guy, "that we shall see a proof of Gilbreath's conjecture in the near future, although the conjecture is probably true." Guy adds that the truth may have nothing to do with the primes as such. Hallard Croft has suggested the conjecture may

```
2  3  5  7  11  13  17  19  23  29  31  37  41  43  47  53  59  61  67  71  73  79  83  89

 1  2  2  4  2  4  2  4  6  2  6  4  2  4  6  6  2  6  4  2  6  4  6

    1  0  2  2  2  2  2  2  4  4  2  2  2  2  0  4  4  2  2  4  2  2

       1  2  0  0  0  0  0  2  0  2  0  0  0  2  4  0  2  0  2  2  0

          1  2  0  0  0  0  2  2  2  2  0  0  2  2  4  2  2  2  0  2

             1  2  0  0  0  2  0  0  0  2  0  2  0  2  2  0  0  2  2

                1  2  0  0  2  2  0  0  2  2  2  2  2  0  2  0  2  0

                   1  2  0  2  0  2  0  2  0  0  0  0  2  2  2  2  2

                      1  2  2  2  2  2  2  2  0  0  0  2  0  0  0  0

                         1  0  0  0  0  0  0  2  0  0  2  2  0  0  0
```

Figure 79

apply to any sequence beginning with 2 and followed by odd numbers that increase at a "reasonable" rate and with gaps of "reasonable" size. If this is the case, Gilbreath's hypothesis may not be as mysterious as it first seems, even though it may be enormously difficult to prove.

One of the most notorious of all unsolved prime conjectures is that there are an infinite number of twin primes. These are pairs of primes that differ by 2. The smallest instances are 3 and 5, 5 and 7, 11 and 13, 17 and 19, 29 and 31, 41 and 43, 59 and 61, and 71 and 73. Many giant examples are known. Until recently the largest example was a pair of 303-digit primes found by Michael A. Penk in 1978. It was surpassed in 1979 when A. O. L. Atkin and Neil W. Rickert found two larger pairs: $694503810 \cdot 2^{2304} \pm 1$ and $1159142985 \cdot 2^{2304} \pm 1$. In the larger twin pair each number begins 4337 . . . , ends with 17760 ± 1 and consists of 703 digits.

The twin-prime conjecture generalizes to prime pairs that differ by any even number n. (Apart from 2, no two primes can have an odd difference because that would make one number even and hence composite.) It can be further generalized to certain finite patterns of numbers separated by specified even differences. For example, the following triplets of primes all fit the pattern k, $k + 2$, and $k + 6$: 5, 7, and 11; 11, 13, and 17; 17, 19, and 23; 41, 43, and 47; and 101, 103, and 107.

It is believed that for any such pattern not forbidden by divisibility considerations there are infinitely many examples. (The pattern k, $k + 2$ and $k + 4$ has only one solution in primes, 3, 5, and 7, because any larger triplet of this pattern would contain a number divisible by 3.) Quartets of the form k, $k + 2$, $k + 6$, and $k + 8$ (the smallest example is 5, 7, 11, and 13) are thought to be infinite. For some patterns no example is known, or only one. R. E. Crandall has called attention to the pattern exhibited by the octet 11, 13, 17, 19, 23, 29, 31, and 37. There are surely other instances of this pattern, but so far none has been found.

The Mersenne numbers—numbers of the form $2^n - 1$, or one less than a power of 2—have fascinated number theorists since classical times, particularly because of their connection with perfect numbers: numbers that are the sum of their divisors, including 1 but not the number itself (6, 28, 496, . . .). If a Mersenne number is prime, it automatically leads to a perfect number by way of Euclid's formula $2^{n-1}(2^n - 1)$, where the number in parentheses is a Mersenne prime.

It is easy to show that a Mersenne number cannot be prime unless the exponent n is prime. If n is prime, will the Mersenne number be prime? The strong law of small numbers suggests it will, because it is true when n equals 2, 3, 5, and 7. The law fails for $n = 11$, however, because $2^{11} - 1$ equals 2047, which equals 23×89. It holds for $n = 13$, $n = 17$, and $n = 19$, then fails again for $n = 23$. From here on successes rapidly become rarer. At the moment only 27 Mersenne primes (hence only 27 perfect numbers) are known. The 27th Mersenne prime, $2^{44497} - 1$, was

discovered in 1979 by a computer program written by David Slowinski
with the assistance of Harry L. Nelson at the Lawrence Livermore Laboratory of the University of California. The number starts 854. . . , ends
. . . 671 and has 13,395 digits. No one knows if the number of Mersenne
primes is infinite, or even if there is a 28th one.

Fermat numbers have the form $2^{2^n} + 1$. For $n = 0$, $n = 1$, $n = 2$, n
$= 3$, and $n = 4$ the number is prime (3, 5, 17, 257, and 65537). Pierre
de Fermat thought all numbers of this form are prime, but he overlooked
the fact that $n = 5$ yields 4294967297, which has the prime factors 641
$\times$ 6700417. No Fermat primes other than the five known to Fermat have
been found, and no one knows whether or not others exist.

Here is a curious pattern involving factorials and primes. Factorial
n, written $n!$, means $1 \times 2 \times 3 \times \ldots \times n$. Note how plus and minus
signs alternate in the following pattern:

$$3! - 2! + 1! = 5$$
$$4! - 3! + 2! - 1! = 19$$
$$5! - 4! + 3! - 2! + 1! = 101$$
$$6! - 5! + 4! - 3! + 2! - 1! = 619$$
$$7! - 6! + 5! - 4! + 3! - 2! + 1! = 4421$$
$$8! - 7! + 6! - 5! + 4! - 3! + 2! - 1! = 35899$$

In each case the number on the right is prime. Alas, the strong law
of small numbers fails on the next step. It yields 326981, the product of
primes 79 and 4139. The next primes result when n equals 10, 15, and
19.

The chart in Figure 80 is formed as follows. We start with 41, then
add 2 to get prime 43. To 43 add 4 to get prime 47. To 47 add 6 to get
prime 53. Continue in this manner, bringing each prime down as the
first number of the next row, and adding numbers from the sequence
2, 4, 6, 8, . . . In every case on the chart the result is a prime. Does this
success continue forever or does it fail at some point?

```
                    EVEN
                    NUMBERS  PRIMES
                       ↓      ↓
                   41 +  2 = 43
                   43 +  4 = 47
                   47 +  6 = 53
                   53 +  8 = 61
                   61 + 10 = 71
                   71 + 12 = 83
                   83 + 14 = 97
                   97 + 16 = 113
                  113 + 18 = 131
                  131 + 20 = 151
                  151 + 22 = 173
                  173 + 24 = 197
                  197 + 26 = 223
                  223 + 28 = 251
                  251 + 30 = 281
                  281 + 32 = 313
                  313 + 34 = 347
                  347 + 36 = 383
                        ⋮
```

Figure 80

The Canadian mathematician Leo Moser constructed the curiosity displayed in Figure 81. A study of the pattern shows that each sequence is formed from the one above it by inserting n, the row number, between all pairs of numbers that add to n. On the right k stands for the number of numbers in each sequence. Note that the first six k numbers are the first six primes. The next k number skips 17, but 19 is a prime. Are all k numbers prime? What is the formula for finding the nth k number?

Figure 81

n	SEQUENCE	K
1	1,1	2
2	1, 2, 1	3
3	1, 3, 2, 3, 1	5
4	1, 4, 3, 2, 3, 4, 1	7
5	1, 5, 4, 3, 5, 2, 5, 3, 4, 5, 1	11
6	1, 6, 5, 4, 3, 5, 2, 5, 3, 4, 5, 6, 1	13
7	1, 7, 6, 5, 4, 7, 3, 5, 7, 2, 7, 5, 3, 7, 4, 5, 6, 7, 1	19

Except for 2, all primes have the form $4k \pm 1$, which means that every prime except 2 is one more or one less than a multiple of 4. (This follows trivially from the fact that every odd number is one more or one less than a multiple of 4.) Write the odd primes in consecutive order, putting the $4k - 1$ primes in the top row and the $4k + 1$ primes under them:

3 7   11 19 23 31 43 47 59 67 71 79 83
5 13 17 29 37 41 53 61 73

At this point the top row is "winning the race." If we continue the two sequences, will the top row always be ahead? You should not waste time trying to settle this empirically, Guy advises, because you have to go a long way before the second row gets ahead, and even then you will not have proved anything. The eminent Cambridge mathematician John E. Littlewood showed that the rows alternately lead infinitely often.

Above 5 all primes have the form $6k \pm 1$. If we race these two "horses," they too change lead infinitely often. Other prime-number races have been investigated, such as the four horses in the $8k \pm 1$, $8k \pm 3$ race. Although it is far from established, most number theorists believe that in all such races, regardless of the number of horses, every horse is ahead infinitely often in the long run.

Primes of the form $4k + 1$ (the bottom row of the $4k \pm 1$ race) can always be expressed as the sum of a unique pair of distinct square numbers. Hence 5 equals $4 + 1$, 13 equals $4 + 9$, and so on. This was proved by Fermat and is known as Fermat's two-square theorem. It is an excellent example of a pattern for which the strong law of small numbers is not deceptive but leads to a genuine theorem. Many ways to prove the theorem have long been known, but in 1977 Loren C. Larson of St. Olaf College in Minnesota published a delightful new proof based on the familiar problem of placing n queens on an n-by-n chessboard so that no queen attacks another.

The figure at the top of Figure 82 shows the smallest solution for the queens problem that displays the following properties: (1) there is a queen on the center square, (2) all other queens are reached from the center by a generalized knight move of m cells in one direction followed by n cells in a direction at right angles to the first (where m and n are distinct integers), and (3) the final pattern has fourfold rotational symmetry (is unchanged by 90-degree rotations). The next-largest solution with all these features is shown at the bottom of the illustration: 13 queens on a 13-by-13 board.

Apart from the center queen, for all such solutions each quadrant of the board obviously must hold the same number of queens. The number therefore will have the form $4k + 1$. Larson shows that solutions of this type can be constructed if and only if the number of queens is a prime of this form.

In all such solutions the board can be divided into identical smaller squares in the manner shown by the slanting lines in Figure 82. If we imagine the board formed into a torus by joining the top and bottom edges and the left and right edges, we see that each board of side p is made up of p tilted squares. Since the board has an area of p^2, the area of each small square is $\sqrt{p}$. Since p is the hypotenuse of a right triangle with sides equal to m and n (the two components of the generalized knight move), it follows from the Pythagorean theorem that p (the area of the square on the hypotenuse) must equal the sum of the squares of m and n. And since p is any prime of the form $4k + 1$, it follows that every such prime is the sum of two distinct squares. I have given Larson's proof, based on earlier work by George Pólya, in highly abbreviated form. For more details see his article "A Theorem about Primes Proved on a Chessboard" (*Mathematics Magazine*, Vol. 50, No. 2, pages 69–74; March 1977).

The fourth and last stanza of Spalding's poem gives a fitting conclusion:

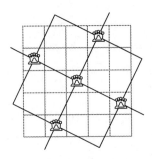

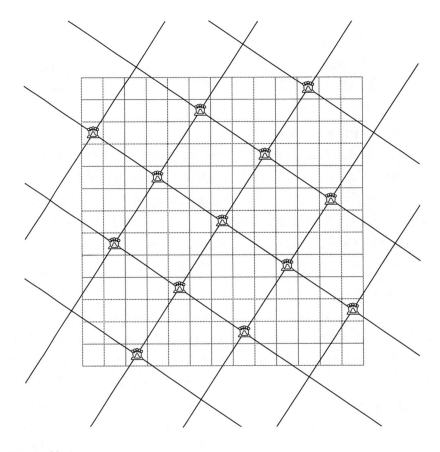

Figure 82

O prime improbable numbers,
Long may formula-hunters
Steam in abstraction, waste
    to skeleton patience:
Stay non-conformist, nuisance,
Phenomena irreducible
To system, sequence, pattern
    or explanation.

Answers

Two questions about prime-number patterns were left unanswered. The first concerned a procedure that seems to generate only primes. Did you recognize this as a clever disguise of Euler's famous prime-generating formula $41 + x^2 + x$? Letting x have integral values starting with 0, the formula generates 40 primes. It fails for $n = 40$, which gives the composite number $1,681 = 41^2$.

Leo Moser's triangle pattern is based on the properties of a sequence known as Farey fractions. It produces a sequence with a prime number of numbers for the first nine rows, but it fails for $n = 10$, which gives a sequence of 33 numbers. If one counts digits instead of numbers, the 10th sequence has 37 digits, a prime, but the next sequence has $57 = 3 \times 19$ digits.

To obtain the k numbers for the nth row, add 1 to the sum of the Euler totients for numbers 1 through n. The Euler totient for a natural number n is the number of natural numbers not greater than n that have no common divisors with n other than 1. For 1 through 10 the Euler totients are 1, 1, 2, 2, 4, 2, 6, 4, 6, and 4. The sum of these numbers is 32. Adding 1 gives the composite number 33 for the 10th row. I do not know if Moser ever published this curiosity.

In 1994 David Slowinski and Paul Gage found the Mersenne prime $2^{858433} - 1$. In 1996 Slowinski and his coworkers at Cary Research found $2^{1257787} - 1$. Later a group of some 700 number crunchers banded together for what they called "The Great Internet Mersenne Prime Search." In 1996 one of its members, Joel Armengaud, discovered the 35th known Mersenne prime, $2^{1398269} - 1$. It has 420,921 digits and is the largest known prime. It provides, of course, a 35th perfect number, also the largest known.

The largest twin prime known to me as I write was found in 1995. It is $242206083 \times 2^{38880}$ plus or minus 1. Each consists of 11,713 digits.

Fermat numbers for $n = 5$ through 9 have now been factored and shown to be composite. The tenth Fermat number, 309 digits long, still seems out of the reach of known methods of factoring large numbers. Incidentally, when written in binary, all Fermat numbers have the form 1 followed by n 0s, with a final 1 at the end. Mersenne numbers in binary consist entirely of 1s.

The chapter mentioned a sequence of eight primes having the form k, $k + 2$, $k + 6$, $k + 8$, $k + 12$, $k + 18$, $k + 20$, $k + 26$, and it stated that the only known instance is 11, 13, 17, 19, 23, 29, 31, 37. John C. Hallyburton, Jr., who works for the Digital Equipment Corporation, found seven other such sequences. The starting numbers of each are

15,760,091
25,658,441
93,625,991
182,403,491
226,449,521
661,972,301
910,935,911

Ken Conrow extended Hallyburton's list of starting numbers for the octets to 49 primes with Hallyburton's six primes as the smallest. His list found all primes below ten billion.

Helen Spalding's poem about primes appeared in the *Guiness Book of Poetry 1958/59*, and in Elizabeth Jenning's *An Anthology of Modern Verse 1940–1960* (Methuen, 1961). I know nothing about Ms. Spalding beyond the fact that she was born in 1920. The poem was sent to me by J. A. Lindon, England, and Philip Gaskell, Glasgow.

References

SEARCHING FOR THE 27TH MERSENNE PRIME. David Slowinski in *Journal of Recreational Mathematics*, Vol. 11, No. 4, pages 458–461; 1978–79.

A SEARCH FOR LARGE TWIN PRIME PAIRS. R. E. Crandell and M. A. Penk in *Mathematics of Computation*. Vol. 33, No. 145, pages 383–388; January 1979.

THE STRONG LAW OF SMALL NUMBERS. Richard Guy in the *American Mathematical Monthly*, Vol. 95, pages 697–712; October 1988.

THE SECOND STRONG LAW OF SMALL NUMBERS. Richard Guy in *Mathematics Magazine*, Vol. 63, pages 3–20; February 1990.

THE EVIDENCE FOR FORTUNE'S CONJECTURE. Solomon W. Golomb in *Mathematics Magazine*, Vol. 54, pages 209–210; September 1991.

PRIME NUMBERS. Second edition. Richard Guy in *Unsolved Problems in Number Theory*. Springer-Verlag, 1994.

13

Checker Recreations, Part I

"The game of draughts we know is peculiarly calculated to fix the attention without straining it. There is a composure and gravity in draughts which insensibly tranquillises the mind."

—James Boswell,
The Life of Samuel Johnson

The quotation is from a section for the year 1756 in which Boswell writes about Johnson's preface to William Payne's *Introduction to the Game of Draughts*, published the same year in London. That book, by a mathematics teacher, was the first in English on the game that in the U.S. is known as checkers. Johnson seldom played the game after leaving college. Boswell expresses regret over it because he thinks checkers playing would have afforded his friend "innocent soothing relief" from periodic bouts of depression.

Nothing is known about the beginnings of checkers, although most game historians now think it originated in southern France sometime in the 12th century. In Britain and the U.S. it is surely the best known of all board games when you consider the number of children who learn to play it and never forget its rules, even though checkers is far below chess in the size of its literature, in the number of adults who become top-level players, and in the public excitement generated by contests for world championship. How many people can name a single checkers expert or tell you who the current world champion is? He is Dr. Marion F. Tinsley, a topologist in the department of mathematics at Florida A. and M. University and probably the greatest checkers player who ever lived.

Rules for chess are now standard throughout the Western world, but not so for checkers. Outside of English-speaking countries there are dozens of regional variations. The version most popular in Europe and Russia, called Polish checkers (except in Poland, where it is called French checkers), is played on a 10-by-10 board, each side starting with 20 men. It is the standard French form of the game. In French Canada the board is even larger: 12-by-12, with 30 pieces to a side. Rules for checkers differ widely around the world. It is curious to note that in all European countries except Britain the pieces are called ladies; only here and in English-speaking countries are they men.

Several consequences follow from the fact that checkers is simpler than chess. One is that a grandmaster checkers player is less likely than his chess counterpart to lose to an inferior by making an error. For checkers buffs this is one of the game's great attractions. They love to quote Edgar Allan Poe's discussion of the two games at the beginning of *The Murders in the Rue Morgue*:

> I will, therefore, take occasion to assert that the higher powers of the reflective intellect are more decidedly and more usefully tasked by the unostentatious game of draughts than

by all the elaborate frivolity of chess. In this latter, where the pieces have different and *bizarre* motions, with various and variable values, what is only complex is mistaken (a not unusual error) for what is profound. The *attention* is here called powerfully into play. If it flag for an instant, an oversight is committed, resulting in injury or defeat. The possible moves being not only manifold but involute, the chances of such oversights are multiplied; and in nine cases out of ten it is the more concentrative rather than the more acute player who conquers. In draughts, on the contrary, where the moves are *unique* and have but little variation, the probabilities of inadvertence are diminished, and the mere attention being left comparatively unemployed, what advantages are obtained by either party are obtained by superior *acumen*.

Tinsley has put it this way: "Playing chess is like looking out over a limitless ocean; playing checkers is like looking into a bottomless well."

Another consequence of the simplicity of checkers is that by 1900 the game's openings had been so completely analyzed that most tournaments ended in draws. To inject more drama into the play Britain introduced (in about 1900) the practice of putting on cards every pair combination of Black's first move and White's response. Before each match a card was chosen at random, and the game had to be played with the specified pair of opening moves. Since each side has a choice of seven moves, there are 49 possible pairs. Two of them (9-14, 21-17, and 10-14, 21-17) were ruled out because they give away a white piece. Later it was found that two more pairs (11-16, 23-19, and 12-16, 23-19) give Black such a strong advantage that they too were discarded, leaving 45 cards.

Standard checkers notation is based on the numbering of squares as shown in Figure 83. For reasons of clarity it is customary in checkers diagrams to reverse the colors of squares and show the pieces on white cells instead of black. Actual play is always on black squares, with the

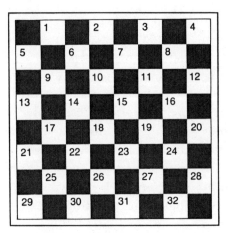

Figure 83

"double corner" at each player's lower right. The players are customarily called Black and White even though the pieces are red and white. Tournament games are now played on green-and-buff boards; black-and-red boards are considered toy-store atrocities. Black always moves first, and games are recorded with Black starting on the low-numbered cells. If you work on any of the recreations in this chapter it is a good idea to label the black squares of your board as shown.

Alas, as decades went by experts soon became so familiar with all variations that follow the two-move openings that "safe" play was adopted and the draws began to pile up again. The British "two-move restriction" was replaced in the U.S. in the mid-1930s by the "three-move restriction," a practice now followed in most checkers tournaments here and in Britain. There are 142 cards, each with a different triplet of the first three moves. Because many of these triplets give an advantage to one side (usually the second player) two games are played with each selection to allow each player the first move.

Without the opening-moves restrictions, a practice known as go-as-you-please play, experts would play nothing but draws. Even with the three-move restriction about 80 percent of all tournament games still end in draws. When an expert does win, it is usually because the loser made a blunder or because the winner managed to keep secret (sometimes for years) a "cook" he had discovered. In checkers a cook is an improvement on standard "book play" that catches an opponent by surprise. Players have traditionally been allowed only five minutes to think before each move and one minute for a capture that can be made only one way. In recent years this practice has been replaced by the use of chess clocks, and players are allowed 30 moves in an hour. When someone springs a new cook, his victim simply does not have enough time to analyze it.

In 1967 the late Walter Hellman, a steelworker in Gary, IN, who was then world champion, defended his title against the U.S. champion, Eugene Frazier. The contest went to 36 games, of which 31 were draws and five were wins by Hellman. Hellman's last win was on a cook. "I had used that cook once before," Hellman told a reporter, "but it had never been published. Frazier had one possible move to thwart the attack, and five minutes doesn't allow much time to figure it out."

A third consequence of the simplicity of checkers is that the best computer programs for checkers play a more formidable game against middle-level players than the best computer programs for chess. Until about 1975 the strongest checkers program was the work of Arthur L. Samuel, a learning program that improves as it plays. After retiring as IBM's director of research, Samuel continued to improve his program at Stanford University's Artificial Intelligence Laboratory. In 1977 a powerful program of the nonlearning type was developed by Eric C. Jensen and Tom R. Truscott, two graduate students at Duke University working under Alan W. Biermann, who teaches artificial intelligence.

Checkers players are ranked on three levels: minor, major, and master. Backers of the Duke program believe it plays initially on a master level. After playing against the program for a while, however, a grand master can discern its weaknesses and begin to exploit them. Its greatest weakness is that it plays without master plans. It does not even follow book moves in opening play, usually scattering its pieces over the board in patterns grand masters consider stupid. Its strength is the incredible speed with which it can analyze all possible moves to much greater depths than a human opponent, and within those depths it never makes a mistake. Chess programs may still be decades away from routinely defeating grandmasters, but the Duke program, Biermann believes, is already "knocking at the door" of the world championship.

Grandmaster checkers players, like their chess counterparts, take a dim view of the quality of computer programs. They all agree with W. Burke Grandjean, secretary of the American Checker Federation, who considers the optimism of the Duke group to be ludicrously naive. Backed by the federation, Tinsley had a standing bet of $5,000 that in a stake match of 20 games he could beat any computer program devised over the next five years. (Readers interested in joining the American Checker Federation and receiving its monthly *Bulletin* can write to Grandjean at 3475 Belmont Avenue, Baton Rouge, LA 70808.) Fidelity Electronics now has on the market Checker Challenger 2, an inexpensive solid-state machine that plays on two levels, and also Checker Challenger 4, which plays on five levels, although its top level is considered below the levels of the Samuel and Duke programs.

In chess it is easy to prove that the "fool's mate," in which the second player checkmates on his second move, is the shortest possible chess game. Surprisingly, the shortest checkers game is not yet known. Until two years ago it was thought to be the 24-move blocked game, the final position of which is shown in Figure 84. There are many sequences of 24 moves that lead to this position, but the position itself is thought to be unique. In the line of play given, every White move is symmetrically opposite (with respect to the board's center) to Black's preceding move. I do not know who first put the play in this symmetrical form. The version I give, worked out by Rudolf Ondrejka of Linwood,

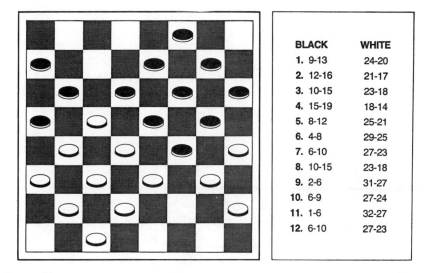

BLACK	WHITE
1. 9-13	24-20
2. 12-16	21-17
3. 10-15	23-18
4. 15-19	18-14
5. 8-12	25-21
6. 4-8	29-25
7. 6-10	27-23
8. 10-15	23-18
9. 2-6	31-27
10. 6-9	27-24
11. 1-6	32-27
12. 6-10	27-23

Figure 84

NJ, begins with the two-move Edinburgh opening. Because 9–13, a favorite first move among tyros, is considered the worst possible start for Black, the symmetrical game is more often started with 10–15, 23–18, an opening known as the Kelso Cross.

Sam Loyd, in his *Cyclopedia of Puzzles* (1914), page 379, using an eccentric notation that incorrectly assumes the board has been rotated 90 degrees, records a nonsymmetrical sequence of moves ending with the same pattern. Loyd states flatly that it is the "shortest possible game." The 24-move blocked game is indeed (as can be proved) the shortest game in which there are no captures. In 1978, however, Alan Malcolm Beckerson, problems editor of *English Draughts Journal*, discovered that White could win on his 10th move (20 moves in all) by capturing all Black's pieces! This is now the shortest checkers game known, although no one has yet proved that no game can be shorter. Beckerson found other 20-movers that capture all the black pieces, as well as some 20-movers that end in blocked games after some captures. The version given in Figure 85, with the board showing the final posi-

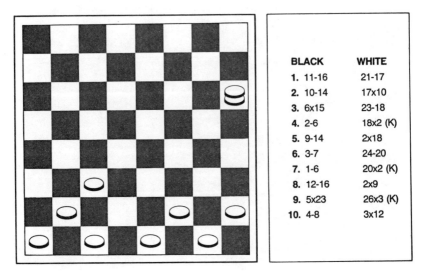

BLACK	WHITE
1. 11-16	21-17
2. 10-14	17x10
3. 6x15	23-18
4. 2-6	18x2 (K)
5. 9-14	2x18
6. 3-7	24-20
7. 1-6	20x2 (K)
8. 12-16	2x9
9. 5x23	26x3 (K)
10. 4-8	3x12

Figure 85

tion, was first published in the British monthly *Games and Puzzles* for March 1978. Its two-move opening is known as the Newcastle.

Many other minimum-move checkers tasks are far from settled. In how few legal moves can a game produce 24 kings? The best-known solution, in 180 moves (90 for each side) by John Harris, appeared in *Journal of Recreational Mathematics* (Vol. 9, No. 1, page 45; 1976). In how few moves (jumps required) can Black and White reverse their initial positions? It takes at least 60 moves for either side, alone on the board, to occupy the opposite starting cells, and so it follows that 2×60, or 120, is an absolute lower bound. A solution in 172 moves is given in a late-19th-century English book, *The Draughts-Player's Guide and Companion*, by Frank Dunne, pages 94–95. At the finish each side has six kings. It seems likely that 172 moves can be considerably lowered.

It is interesting to try this problem on smaller boards. The 3-by-3 is trivial, but the 4-by-4 presents a pleasant puzzle. Starting as is shown in Figure 86, the task is to interchange the two sides in a minimum number of legal moves. Captures are of course compulsory. At the finish all

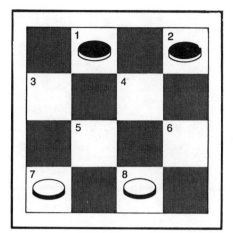

Figure 86

four pieces will necessarily be kings. Incidentally, five moves are needed for the shortest game on this miniboard. If both sides play to win and follow their best strategy, the game is a draw.

As with chess, endless ways of playing checkers have been proposed by varying the shape of the board, the starting position, the rules and so on. A privately published French book, *Les jeux de dames non orthodoxes et autres jeux à pions*, by Joseph Boyer and Vern R. Parton, gives more than 100 such variants. Some are played on triangular or hexagonal tessellations and some on three-dimensional boards; some mix chess pieces with checkers, and some allow three or four players to compete at once. As one would imagine, it is hard to draw a line between a game similar enough to checkers to be called a variant and one so different from checkers that it is best regarded as another game altogether. The so-called Turkish checkers, for example, has almost no resemblance to checkers except that it is played on an 8-by-8 board with counters of two colors. One simple way to vary standard checkers is to start with the men positioned as is shown in Figure 87. All checkers rules hold. The opening moves quickly lead to patterns not encountered in orthodox games.

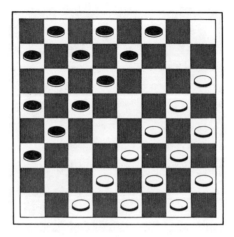

Figure 87

The most eccentric variation of checkers, about which one would like to know more, is "supercheckers," invented by Charles Fort, the Bronx collector of science anomalies who had a marked influence on science fiction as well as on the current epidemic of interest in the paranormal. According to Fort's biographer Damon Knight (*Charles Fort*, Doubleday, 1970), supercheckers was played "with armies of men on a board with thousands of squares. Fort used bits of cardboard with carpet-tack handles for the men, and a piece of checkered cloth for the board."

The two players start with their forces in any agreed-on formation that has a space between the two armies. If a player moved only one man at a time a game might last for weeks, and so Fort allowed for movements en masse. Here is how he put it in a letter: "Let A start out, moving until B tells him to stop—say a hundred moves. Then B makes a hundred moves. A may want to make another hundred moves, but B, sizing up the situation, tells him to stop, say at thirty. Then perhaps occurs 'fighting,' at close quarters, one move at a time, as in ordinary checkers. But, at any time, if either player wants to make a 'mass movement,' that is a matter of obtaining permission from his opponent."

A game usually lasted all night. In 1930 Fort wrote to Tiffany Thayer, who edited the first Fortean magazine, *Doubt:* "Supercheckers is going to be a great success. I have met four more people who consider it preposterous."

In Britain and the U.S. the most popular variant of checkers is "giveaway." It differs from the standard game only in that the object is to be the first to *lose* all one's men. In Dunne's book cited above, pages 91–92, there is a fantastic giveaway "sucker bet," presumably devised by British checkers hustlers. White begins with his 12 men in the usual starting position. Black has only a king on cell 7. Black wins if he loses his king. White wins if he loses all 12 men. Dunne shows how White can always win and gives three similar wagers in which Black begins with a single uncrowned piece on cell 1, cell 4, or cell 5.

Among hundreds of hustler wagers, one of the best begins with the position shown in Figure 88. (I am indebted to Mel Stover for passing it along.) It is Black's turn. White wagers that Black will not be able to crown the piece he moves first. Clearly Black should not move the piece on cell 21 because he would lose it immediately, so that the question is

Figure 88

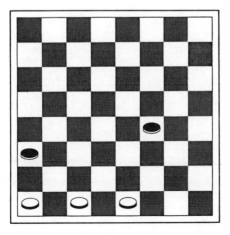

whether Black can move the piece on cell 19 and advance it to his king's row. The more you study the pattern, the more obvious it seems that Black can win the bet easily. Nevertheless, White wins. It is an amusing bet to make with friends.

One final problem. It is widely believed two kings can always win against one king, but that is not invariably true. See if you can place two white kings and one black king on the board in such a way that when it is White's turn, Black can force a draw. We ignore the trivial case in which black is between two white pieces along a diagonal.

Answers

The first problem was to exchange the position of the two black and the two white pieces on a 4-by-4 minicheckerboard, making the fewest possible moves. The minimum number of moves needed to solve this minicheckers puzzle is 16. Numbering the black squares of the board 1 through 8, as was shown, the first four moves must be 2-4, 8-5, 4-6, and 5-4. The fifth move may be 1-3 or 6-8, with many variations thereafter. A typical sequence of the last 12 moves is 1-3, 4-1, 6-8, 7-5, 8-6, 5-4, 3-5, 4-2, 5-7, 1-3, 6-8, and 3-1.

White can win the checkers hustler's wager as follows:

Black	White
19-24	29-25
24-28	30-26
21-30	31-27
30-32	

At this point the game is over, and although Black has won the game, he has failed to crown the piece he moved first, thereby losing the wager.

Finally, Figure 89 shows a checkers position in which White has two kings against Black's one king and is the next to move. By playing properly Black can force a draw. It is the only position in which one king can draw against two kings, aside from familiar traps in which Black slides between two White kings on a diagonal. If neither White king is at the board's edge, Black forces a capture of one of the White kings. The trap also works if the White kings are at a single corner, say at 30 and 21, and Black is on 22.

Herschel F. Smith pointed out that a Black king on 2 also wins against White kings on 6 and 7, but of course there is no way this joke position can occur with legal moves.

Figure 89

References

SOME STUDIES IN MACHINE LEARNING USING THE GAME OF CHECKERS. A. L. Samuel in *IBM Journal of Research and Development*, Vol. 3, pages 210–229; July 1959.

SOME STUDIES IN MACHINE LEARNING USING THE GAME OF CHECKERS, II: RE-
CENT PROGRESS. A. L. Samuel in IBM *Journal of Research and Development*, Vol. 11, pages 601–617; November 1967.

THE COMPLEXITY OF CHECKERS ON AN N × N BOARD—PRELIMINARY REPORT.
A. S. Fraenkel, M. R. Garey, D. S. Johnson and Y. Yesha in *19th Annual
Symposium on the Foundations of Computer Science*. Institute of Electrical
and Electronics Engineers, 1978.

A PROGRAM THAT PLAYS CHECKERS CAN OFTEN STAY ONE JUMP AHEAD. A. K.
Dewdney, in *Scientific American*, pages 14–27; July 1984. The article is
reprinted in Dewdney's *The Armchair Universe*. W. H. Freeman, 1988.

A CHARLES FORT INVENTION: SUPER-CHECKERS. Charles Fort in *The INFO
Journal*, pages 24ff; June 1990.

THE CHECKER CHALLENGER. Ivars Peterson in *Science News*, Vol. 140, pages
40–41; July 1991.

14

Checker
Recreations,
Part II

Since writing the previous chapter in 1980 I have encountered so much new material about checkers that I decided to patch it together in a new chapter rather than try to force it into the older one.

In its August 1980 edition *Scientific American* published the following letter from Marion Tinsley, of Tallahassee, FL, then the World Checkers Champion:

> Sirs:
>
> I have a few comments about Martin Gardner's January

article on checkers. He writes: "When an expert does win, it is usually because the loser made a blunder or because the winner managed to keep secret . . . a 'cook' he had discovered." This is completely misleading. Certainly blunders occur and cooks are pulled. Knowledge *is* important in checkers. However, the ability to see deeply into the game is far more important. In his two world-title matches with Asa Long (1948 and 1961) Walter Hellman sprang a great number of cooks from his vast storehouse of knowledge. Yet Long, because of his tremendous analytical ability, made the wins so difficult that Hellman could not find them. Wins between experts tend to be very narrow indeed. In 1922 Long won his first national tournament at the age of 18, besting players who greatly surpassed him in knowledge. He did it on ability, not cooks. One further illustration. In 1876, at the age of 19, R. D. Yates won the world title from James Wyllie. Wyllie had been a great scholar of the game for 40 years and had introduced many important openings and developments. Yet Yates beat him. His games were not a testimony to his knowledge but a monument to his great ability to see deeply into the game.

Finally, a remark about checkers-playing computer programs. I have seen games played by most of them, including six games played by the Duke program. They all play at the very weak amateur level. The programs may indeed consider a lot of moves and positions, but one thing is certain. They do not see much! Nevertheless, for 20 years claims have been made repeatedly that there exist programs playing at the master level. It is because of exasperation with such false and aggravating claims that the wager has been made. We are not a fraternity of gamblers. The idea of a stake challenge, however, has become accepted as the only way to effectively expose fakery. Perhaps someday the programmers will have a real breakthrough. But until then let them behave like true scientists and refrain from undue boasting about their offspring.

The 120-move way to interchange the two sides with legal moves, both sides cooperating, has an interesting history. The problem was first proposed by a Dr. Brown in England's *Gentleman's Journal* (September 1872), who gave a solution in 172 moves. Jumps are, of course, compulsory, so in any solution no possibility of a jump can arise. A Mr. Harber, in a four-part article in the *Weekly Times* of Melbourne, Australia (June 19 and 26, and July 3 and 10, 1968), reduced the number of moves to 120! Brown had wasted many moves by pushing kings backward and forward to make "traffic lanes" for other pieces. In the final article of the series—the series was titled "The Interchange"—Harber included a proof that 120 moves was minimal.

For the 14-move checker game with no captures, Alan Malcolm Beckerson, of London, found 28 possible final positions, of which only two (including the one I gave) show symmetry. In 16 of them, were it not Black's turn, White would be able to make one move more.

I carelessly said it could be proved that no game shorter than 24 moves, with no captures, was possible. I was wrong. In 1963 Beckerson composed several no-capture games that end after the 21st move. One was published in *Games and Puzzles* (June 1976), and recognized as the shortest checkers game in the 24th edition of the *Guinness Book of Records*. One of five such games is shown in Figure 90.

Checkers on a 4-by-4 board is a draw when played rationally. A large part of the complete game tree was given by A. K. Dewdney in his *Scientific American* column listed as a reference for the previous chapter.

Dewdney is convinced that the 6-by-6 game is "very likely" a draw, and the standard 8-by-8 "probably" a draw. In a much earlier *Scientific American* column, reprinted as Chapter 8 in *The Unexpected Hanging and Other Mathematical Diversions*, I gave reasons for being almost certain that the 4-by-4 game is a draw.

What about the 5-by-5, each side starting with three checkers on its first row? The surprising answer, as I explain in the chapter cited above, is that the first player has a sure win! The board's lack of a double

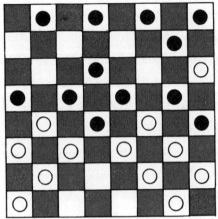

FINAL POSITION

	BLACK	WHITE
1.	12-16	22-17
2.	16-20	23-19
3.	11-15	19-16
4.	9-14	16-12
5.	14-18	26-22
6.	5-9	31-26
7.	9-14	26-23
8.	6-9	23-19
9.	9-13	30-26
10.	7-11	26-23
11.	11-16	

Figure 90

corner, where a king can safely shuttle back and forth, is responsible for eliminating a forced draw.

Gilbert Chesterton, in the second chapter of *Orthodoxy*, his great book of Christian apologetics, had this comment on Poe's preference for checkers over chess:

> Moreover, it is worthy of remark that when a poet really was morbid it was commonly because he had some weak spot of rationality on his brain. Poe, for instance, really was morbid; not because he was poetical, but because he was specially analytical. Even chess was too poetical for him; he disliked chess because it was full of knights and castles, like a poem. He avowedly preferred the black discs of draughts, because they were more like the mere black dots on a diagram.

A good checkers player is seldom interested in chess, and good chess players usually have a similar disinterest in checkers, but there are at least three notable exceptions. Harry Nelson Pillsbury, a chess grand-master, was also a master at checkers. Newell Banks was both a check-ers master and a top chess player. The third person was Irving Chernev, a master at both games, and a popular writer of books about chess. In *Chess Life and Review* (September 1979) he had this to say:

> In fact I gave up chess for five years, in my twenties, to study checkers. Early in my youth I was beaten badly, and I decided that nobody would ever beat me that way again. They might beat me, but not that way.
>
> I was interested in seeing how great masters played, and I discovered that there was a great literature on checkers, and that it could be a great game. There's a lot of beauty and science in it. And so I've decided to write a book on check-ers, and to give it all that I've discovered in the intervening.

I once invented a checkers-type game called Solomon. Played on a board based on the Star of David, its rules are the same as in checkers. It is not known which side can win in rational play, or if Solomon is a draw, although the game is simple enough to be solvable by a computer program. An interesting feature of the game is that two kings can always defeat a single king, although the strategy for doing so is harder to discover than the strategy by which two checkers kings defeat a single king moving back and forth in a double corner. The game is available from Kadon Enterprises, 1227 Lorene Drive, Pasadena, MD 21122.

It has long been noticed that two games of checkers can be played simultaneously on the same board, one game played on black squares, the other played on white squares.

What is the maximum number of kings that can be jumped by a king of opposite color? The answer is nine, arranged in a 3-by-3 square formation.

Two kings can usually defeat a single king. Not so well known is that three kings can also defeat two opposing kings even when the two are in double corners. In general, the best way for three kings to win against two kings is to force an exchange, leaving two against one.

A Braille board and men for blind checker players has been marketed. The squares are indented to keep the men in place. Checkers of one color are square, those of the other color are round.

Reader Abe Schwartz called my attention to the fact that in the miniature 4-by-4 game two kings cannot defeat a single king in the double corner. The board must be at least 6-by-6 for the familiar strategy to work. Schwartz also found that three kings against two kings, each in a double corner, can win on the standard 8-by-8 board, but the game is a draw on all larger boards.

Ike Kisch, then secretary emeritus of The American International Checkers Society, wrote to say that the term "Polish Checkers," played on a 10-by-10 board, is now called "International Checkers." According to Kisch, the name Polish Checkers originated in France about 1750 when a Polish gentleman introduced the 100-squares board. It became popular in France, and quickly spread to other countries, becoming especially popular in Russia and Holland.

For readers familiar with computer complexity, it has been shown that checkers generalized to a $2n$-by-$2n$ board, like generalized go, is "P-space hard." This implies, for example, that other games like chess and go, when generalized to $n \times n$ boards, can be simulated by equivalent checker positions on boards whose size is a polynomial function of n. The proof that checkers is P-space hard was made by Aviezri Fraenkel, in Israel, in collaboration with Michael Garey and David Johnson, of Bell Labs.

John Rogers (1829–1904) was a famous early American sculptor whose plaster-of-Paris statuettes of familiar scenes made him the three-dimensional Norman Rockwell of his time. One of his most popular works was "Checkers Up at the Farm." Five thousand copies were sold throughout the nation. In 1979 the John Rogers Commemorative Society made and sold a bronze copy of this work in an edition limited to 650. (See Figure 91.) The young man is pointing gleefully at his winning move. Original plaster-of-Paris copies of Rogers' many statuettes often turn up in antique stores and shows, usually in chipped condi-

Figure 91

CHECKERS
UP AT THE FARM

tion, where they sell from $500 to more than $1,000 depending on their condition.

Founders of artificial intelligence have been notoriously wrong in predicting when a computer chess program would defeat all grandmasters to become world champion. Similar over-optimistic prophecies have been made about checkers. Richard Bellman of the Rand Corporation, for example, published an article "On the Application of Dynamic Programming to the Determination of Optimal Play in Chess and Checkers," in the *Proceedings of the National Academy of Sciences* (Vol. 53, February 1965, pages 244–247). In this paper he wrote: "With bigger computers . . . it seems safe to predict that within ten years checkers will be a completely decidable game."

More than thirty years have passed since Bellman's rash prophecy, and checkers is still far from having been decided, although checkers programs are rapidly improving. At the time I write (1996) there are several strong programs commercially available but the best is CHINOOK, developed at the University of Alberta, Edmonton, Alberta, Canada, by three computer scientists: Jonathan Schaeffer, Robert Lake, and Paul Lu, with the assistance of two checkers experts, Martin Bryant and Norman Treloar. The story was dramatically told in 1996 in "CHINOOK, the World Man–Machine Checkers Champion," by Schaeffer, Lake, Lu, and Bryant. A nontechnical book about CHINOOK, by Schaeffer, is scheduled for publication by Springer-Verlag in 1997.

Tinsley first played CHINOOK in a 1990 exhibition match, winning one game, losing none, and drawing 13. He and the program were officially matched in London in 1992. Tinsley won 4 games, lost 2, and drew 33. The losses were only the sixth and seventh defeats by Tinsley in 42 years!

A rematch was held in 1994. CHINOOK had greatly improved, with scores of new and secret "cooks" of openings, and an ability to search all branches of the game tree for at least 21 moves! The first six games were draws. Tinsley resigned to have health tests made. He was diagnosed with cancer, and died in 1995, an undefeated champion. His resignation gave the world championship to CHINOOK, leaving open the question of whether Tinsley or CHINOOK was the better player.

CHINOOK retained its title by drawing a match with master Don Lafferty. In 1995 it won a rematch with Lafferty by winning one game, losing none, and drawing 31.

The present human world champion is Ron King. So far he has not had an official match against CHINOOK, but Schaeffer and his associates are confident that since Tinsley's death there is no checkers player around capable of defeating CHINOOK. The world's four top players are rated as follows: CHINOOK 2712, Ron King 2632, Asa Long 2631, and Don Lafferty 2625.

CHINOOK improves almost daily, and its makers are actually hoping to achieve the goal of "solving" checkers by improving their program until it plays a perfect game.

A truly fantastic checkers bet was contributed by Mel Stover to *Recreational Mathematics Magazine* (April 1961), a forerunner of the *Journal of Recreational Mathematics*. With Stover's permission, here is how he dramatized the bet:

> To follow the different possibilities in this clever production, let's visit the "Mythical Chess and Checker Club". Joe Kalyika was absorbed in a game of checkers with his friend Sam Palooka. Except for a lone kibitzer the club was empty.
>
> Sherwin Betts, a stranger to the checker players, was well known to the pool hall gentry as a shrewd operator who made a good living out of gambling. He had already classified the two as mediocre players and he had noticed that Joe Kalyika had a large measure of that ostentatious smugness known to grifters the world over as the mark of the mark.

Sam had moved 7–10 and Joe claimed the game. (See
Figure 92.) "I go here and you go there and I go here and you
go either there or there and I go here and win.

(translation)	22–17	21–25
	17–13	25–29 or 30
	18–14 and White wins	

Betts interrupted this demonstration, "I know it's none
of my business but Black should draw that game every day in
the week."

"Are you out of your cotton-picking mind? That is a clear-
cut win for White," shouted Joe.

"I can see that a beginner might think so," said Betts in
a tone calculated to irritate.

"Beginner, am I. Put up some money and take the Black
and I'll demonstrate it move by move."

Betts placed a five dollar bill on the table and occupied
Palooka's empty chair. This time the game proceeded:

22–17	21–25
17–13	10–14
18–9	25–30
9–6	30–26
6–2	26–23 and Black draws

Kalyika was silent and Betts made no attempt to pick up
the money. "I don't like to take a person's money without

Figure 92

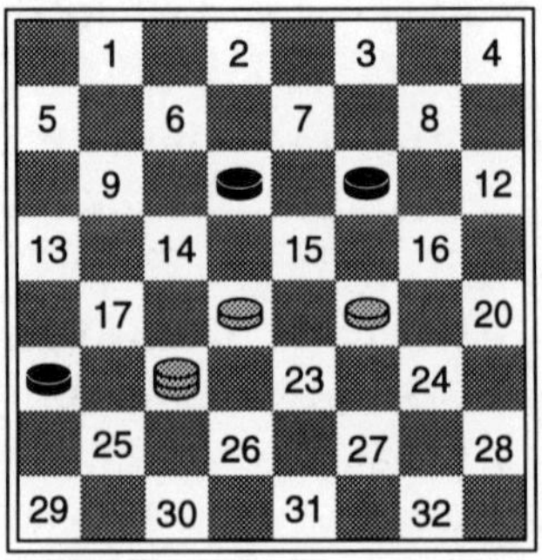

giving him a chance to win it back. So for ten dollars I'll let you take either side. If you take the White you must win. If you take the Black you must draw." Kalyika placed a tenspot beside the two fives and sat down at the seat that Betts had just vacated. Betts moved 22-17 and the game went:

	21-25
17-21	25-30
18-14	10-17
21-14	30-26
14-18	and White wins

Betts' smile infuriated Joe. Kalyika replaced the checkers and sat down at the white side of the board. "I guess it is my turn at the White," said Kalyika as he placed twenty dollars on the table. Betts, still smiling, again took the black men.

22-17	21-25
17-21	10-14
18-9	25-30
9-6	30-26
6-2	26-23 and Black draws

"Want to try again?" asked Betts. Kalyika said nothing but a clue to his thoughts was visible in the puce tone which tinged his ruddy complexion. Once again they changed places. By now the wager was forty dollars. The moves:

22-17	21-25
17-21	10-14
18-9	25-30
21-25	30-21
9-6	21-17
6-2	17-14
2-7	and White wins

"Perhaps if you took the Black and we bet eighty dollars," said Betts as he turned around but Kalyika had already gone.

Stuffing the bills into his wallet Sherwin Betts sadly shook his head and slowly walked away.

References

CHINOOK, THE WORLD MAN-MACHINE CHECKERS CHAMPION. Jonathan Schaeffer, Robert Lake, Paul Lu, and Martin Bryant, in *AI Magazine*, pages 21–29; Spring 1996.

15

Modulo Arithmetic and Hummer's Wicked Witch

There was a young fellow named
                Ben
Who could only count modulo
                ten.
        He said, "When I go
        Past my last little toe,
I shall have to start over again."

Congruence theory (sometimes called modular arithmetic) is based on principles as old as arithmetic, but it was the German "prince of mathematicians," Karl Friedrich Gauss (he has been called the greatest mathematician who ever lived), who pulled them all together and unified them with a notation so compact and powerful that it is hard to imagine how number theory could have advanced without it. The son of an uneducated bricklayer, Gauss was a child prodigy whose most influential book, *Disquisitiones Arithmeticae*, was published by himself in 1801 when he was 24. He had written it four years earlier. It was this book that introduced the concept of number congruence.

Gauss defined two integers a and b to be congruent for a modulus m (modulus is Latin for a small measure) if their difference is divisible by a nonzero integer m. To say the same thing another way, two integers are congruent modulo m if they have the same remainder when they are divided by m. Gauss symbolized congruence by three short parallel lines, a symbol still used today: $a \equiv b$ (mod m). Incongruence is indicated like this: $a \not\equiv b$ (mod m).

For example, 17 and 52 are congruent modulo 7 because each has a remainder of 3 when it is divided by 7. Expressed the other way, $52 - 17 = 35$, which is 7×5. If we call the multiplier k (in this instance k is 5) and let a be the larger integer, then $b = a + km$, where m is the modulus and k is some integer. Many of the rules of ordinary arithmetic and algebra (such as addition, subtraction, and multiplication) apply to the manipulation of congruences.

Remainders are called residues, and for every modulus m there are m "residue classes." The smallest modulus, 2, distinguishes even and odd numbers. All even numbers are congruent to 0 (mod 2) and have the infinite residue class . . . $-4, -2, 0, 2, 4, \ldots$ All odd numbers are congruent to 1 (mod 2) and have the infinite residue class . . . $-3, -1, 1, 3, 5, \ldots$. For $m = 3$ the residues are 0, 1, and 2. There are three infinite classes (mod 3) and so on for higher values of m.

As Gauss made clear, his congruence algebra provided simple proofs for various rules that determine whether a number is divisible by a given number. (From here on "number" will mean "integer.") Thus n is divisible by 3 if and only if the sum of its digits is congruent to 0 (mod 3). Similarly n is congruent to 0 (mod 9) if and only if the sum of its digits is congruent to 0 (mod 9). A number n is congruent to 0 (mod 4) if and only if its last two digits form a number congruent to 0 (mod 4), and n is congruent to 0 (mod 8) if and only if its last three digits form a number congruent to 0 (mod 8). A number is congruent to 0 (mod 11) if and only if the difference between the sum of its digits in even positions and the sum of its digits in odd positions is congruent to 0 (mod 11).

Congruence algebra led to important theorems about prime numbers and also simplified proving them. For example, Fermat's "little theorem," which is useful in testing for primality, states that if a number a is raised to the power of $(p - 1)$, where p is a prime that does not divide a, then when the result is divided by p, the remainder is always 1. In Gauss's terminology, $a^{(p-1)}$ is congruent to 1 (mod p). Thus a number can be raised to the power of one less than a prime so large that the result can have billions of digits and be far beyond the ability of computers to calculate, yet we know that if we subtract 1 from this unprintable monster, we shall have a number that is a multiple of the prime.

Another famous result related to Fermat's little theorem is known as Wilson's theorem. If you multiply consecutive numbers starting with 1 and stop at any number immediately preceding a prime, the product obviously is divisible by any number up to p but not by p itself. If you add 1 to the product, however, lo and behold the result becomes a multiple of p. For example, 1 times 2 times 3 times 4 is equal to 24, which is not divisible by the next number, 5, a prime. But 24 plus 1 is equal to 25, which *is* a multiple of 5. Using factorial and congruence signs, Wilson's theorem is $(p - 1)! + 1 \equiv 0 \pmod{p}$.

The theorem was known to Leibniz but was rediscovered by a British scholar named John Wilson. Edward Waring credited it to him in a 1770 algebra book and remarked that the theorem would be extremely difficult to prove because mathematicians had no good notation for primes. When Gauss was told this, he commented that for such proofs one needs not *notationes* (notations) but *notiones* (notions). Wilson's theorem is a marvelous criterion for primality, but unfortunately it is of no use in computer searches for big primes.

Thousands of basic theorems in number theory are compactly expressed and their proofs made easy and elegant by modular theory, and endless puzzles have been based on such theorems. For example, suppose a manufacturer of dice ships his product to wholesalers in large cubical boxes. A wholesaler removes one row of dice from the cubical array to test them for possible flaws, and during the tests these dice are destroyed. The remaining dice are packed into small boxes, six to a box. How many dice are left over? Surprisingly, regardless of the size of the original box none are left over. This follows from the congruence theorem $n^3 - n \equiv 0$ (mod 6).

Here is a problem that demonstrates the power of congruence algebra to provide solutions. (I found it in Allan Gottlieb's "Puzzle Corner" in *Technology Review* for May 1978.) You want to prove the curious theorem that every integer n has some multiple that consists of a string of 1s followed by a string of 0s. How can you go about it? One way is to list n "rep unit" numbers starting with 1, 11, 111, 1111 up through n such numbers. The number of possible remainders when any number is divisible by n is obviously n. To our list of n rep-unit numbers we add one more. On the pigeonhole principle at least two numbers on this list must have the same remainder and therefore be congruent modulo n. Now, the difference between any two numbers that are congruent modulo n is congruent to 0 (mod n), which means that the difference is a multiple of n. Therefore we subtract the smaller of the pair of congruent rep-unit numbers from the larger, and the result will be a number of the form we seek.

To see better how this works, let us find a number of the form 111 . . . 0 . . . that is a multiple of 7. The first eight rep-unit numbers are 1, 11, 111, 1111, 11111, 111111, 1111111, and 11111111. Their residues (mod 7) are respectively 1, 4, 6, 5, 2, 0, 1, and 4. Since there are eight numbers, we must have at least two numbers with the same residues (mod 7). In this instance there are two such pairs. The smallest pair is 1 and 1111111. The difference is 1111110, or 7 × 158730. It is the smallest number of the form we seek.

Measurements of time in most cultures are made in modular systems. We measure hours by a mod-12 arithmetic. If it is 3:00 now and we want to know what time it will be 1,000 hours from now, we simply add 1,000 to 3, then divide 1,003 by 12. The residue, 7:00, is our answer. The clock is such a familiar model of a modular system that when schoolteachers introduce number congruences, they like to

THE FAR SIDE   By GARY LARSON

Early checkers

call it "clock arithmetic." The U.S. armed forces use a mod-24 clock. Days of the week conform to mod-7 arithmetic, the months of the year to mod 12, and the years of the century to mod 100.

Many problems about the calendar yield readily to congruence formulas. Gauss himself gave algorithms for determining the day of the week when one is given the year and the day of the month, and also algorithms for calculating the date of Easter. According to the Gospels, the resurrection of Jesus took place on a Sunday morning during the Jewish Passover week, celebrated after the first full moon of spring. The early Christians wanted to keep the symbolic connection between the Passover sacrifice and the sacrifice of Christ, and so it was decided at the First Council of Nicaea (A.D. 325) that Easter would be the first Sunday after the first full moon after the vernal equinox. Unfortunately the old Julian calendar made the year slightly longer than it actually is, so that the date of the vernal equinox was creeping slowly backward from March 21 toward April.

When Pope Gregory XIII introduced the present calendar in 1582, he did so mainly to restore Easter to spring. It is a sad commentary on the Middle Ages that calculating the exact dates of Easter was then one of the most important of all applications of mathematics to nature.

Gauss's algorithms for determining Easter dates in both the Julian and the Gregorian calendars are complicated, and they have to be patched by special rules to take care of exceptions. If we limit our concern to the years from 1900 to 2099 inclusive, however, there is a straightforward procedure, with no exceptions, that was devised by Thomas H. O'Beirne of Glasgow and first published in his paper "The Regularity of Easter" (*Bulletin of the Institute of Mathematics and Its Applications*, Vol. 2, No. 2, pages 46–49; April 1966). O'Beirne found he could memorize his procedure and as a party stunt give the date of Easter for any year during the relevant period by making all the calculations mentally.

O'Beirne's algorithm is summarized in Figure 93. Easter always

---

1. Call the year Y. Subtract 1900 from Y and call the difference N.
2. Divide N by 19. Call the remainder A.
3. Divide (7A + 1) by 19. Ignore the remainder and call the quotient B.
4. Divide (11A + 4 − B) by 29. Call the remainder M.
5. Divide N by 4. Ignore the remainder and call the quotient Q.
6. Divide (N + Q + 31 − M) by 7. Call the remainder W.
7. The date of Easter is 25 − M − W. If the result is positive, the month is April. If it is negative, the month is March (interpreting 0 as March 31, −1 as March 30, −2 as March 29 and so on to −9 for March 22.

Figure 93

falls in March or April. The earliest possible date is March 22. It last happened in 1818 (when it fell on a full-moon day), and it will not happen again until 2285. The latest possible date is April 25. It last happened in 1943, and it will not happen again until 2038. You might like to test O'Beirne's procedure to see that it correctly gives April 6 for Easter in 1980, April 19 for 1981, and April 11 for 1982. April 19 is the most frequent of all Easter dates, with April 18 running a close second.

Countless magic tricks, particularly with numbers and playing cards, are based on congruences, and many have been described in my *Scientific American* columns. A trick I have not discussed earlier depends on the fact that the sum of all the values of the 52 cards in a deck is $364 \equiv 0 \pmod{13}$. (Jacks count as 11, queens as 12, and kings as 13.) Let someone shuffle the deck, then remove a card without anyone's seeing its face. After dealing just once through the deck of 51 cards, looking at the face of each card, you correctly name the card that was removed.

Magicians have devised many algorithms for this trick, but the following one seems to me the easiest. As you deal the cards keep in your head a running total of the values but cast out 13 as you go along. In other words, whenever the total goes above 13, subtract 13 and keep in mind only the difference. The task is greatly simplified by two rules:

1. Ignore all kings. Their value, 13, is congruent to 0 (mod 13); therefore they do not alter the number you keep in mind.

2. For 10s, jacks, and queens, instead of adding 10, 11, and 12, subtract 3, 2, or 1 respectively. This reflects the fact that in the mod-13 system 10 is congruent to −3, 11 is congruent to −2, and 12 is congruent to −1.

After the last card is turned subtract the number in your head from 13 to get the value of the missing card. If the result is 0, the card is a king.

How do you know the suit? A good procedure is to use your feet for secret calculating in mod-2 arithmetic. Start with both feet flat on the floor. For each spade raise or lower your left heel. For each club raise or lower your right heel. For each heart alter the positions of both feet simultaneously. Ignore all diamonds. After the deal your feet indicate the suit of the missing card as follows:

—If only the left heel is up, the card is a spade.
—If only the right heel is up, the card is a club.
—If both heels are up, it is a heart.
—If both heels are down, it is a diamond.

After some practice it is surprising how quickly you can deal through the deck and name the missing card.

Robert Hummer, a magician, has been unusually productive in inventing mathematical tricks, and many of his creations are based on mod-2, or odd–even, principles. I give here for the first time a set of mysterious fortunetelling cards that is one of Hummer's most ingenious ideas.

First you must make a set of the seven cards shown in Figure 94. Photocopy them, paste them on a sheet of cardboard, and cut them out. Here is how they are used.

You are allowed to ask the Wicked Witch of the West only one question a day. Of course, you may experiment with more questions if

Figure 94

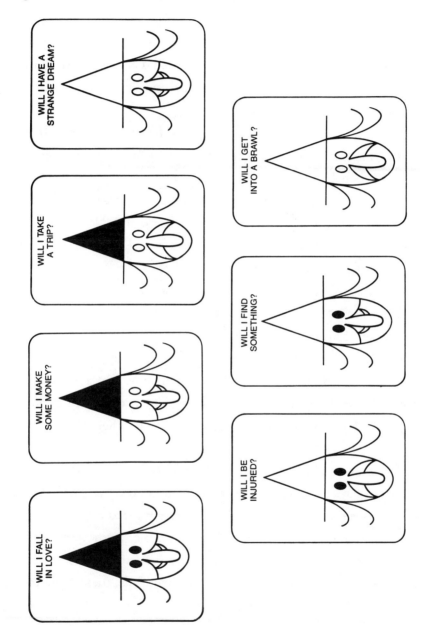

you like, but the answers are not guaranteed to be trustworthy. Each answer applies only to a period of seven days following the day the question is asked. Select the card with the desired question and put it aside. Shuffle the remaining six cards and hold them face down in one hand. Wave your other hand over the packet and slowly pronounce the mystic precognitive mantra "Puthoffa Targu."

From the top of the packet remove the first pair of cards. If the colors of the hats match, put the cards aside to form a pile. Discard them if the hats fail to match. Repeat with the next pair. If the colors of the hats match, put the pair on top of the pile. Otherwise discard them. Check the remaining pair and repeat the procedure. Now count the number of matching pairs. The number will be 0, 1, 2, or 3. Write this down as the first digit of a three-digit number.

Assemble the six cards, shuffle, pronounce the mystic mantra, and repeat the procedure, except this time look for matching eyes. Record the number of matching pairs as the second digit of your number.

Shuffle the six cards for the third and last time, say the mantra and go through the packet by pairs as before. This time look for matching expressions (smile or frown). The matching pairs are counted—remember, you count pairs, not single cards—to get the last digit of your number.

Find your number in Figure 95 and read the answer. Even though the digits of your number were randomly obtained, you will find a specific answer that applies only to the question asked.

If you want to ask the Wicked Witch a yes–no question that is not on any card, you may do so, but now you must use all seven cards. Follow the same procedure, looking first at the hats, then at the eyes, and then at the expression. This time, however, you must form two piles, one of matching pairs and one of nonmatching pairs. Ignore the last card. Subtract the number of pairs in the smaller pile from the number of pairs in the larger and record the number. After three trials

Figure 95

```
000 You will dream about a relative.
001 You will have an argument on the telephone.
002 You will dream about elephants.
003 You will exchange angry words with a plumber.
010 You will find a lost ring.
011 Something you say will harm you.
012 You will find the weather abominable.
013 Be alert for an injury to your foot.
020 You will dream about an old friend.
021 Yes, but it will be a fight you did not start.
022 You will dream about an airplane.
023 Not if you can control your temper.
030 You will find a coin on the street.
031 Only a slight nick while shaving your face or your legs.
032 You will find a lost object in the pocket of an old bathrobe.
033 No, but you will injure someone else.
100 No, because you know counterfeiting is illegal.
101 You will make a trip to the liquor store.
102 Just the usual amount.
103 You will make a short journey south.
110 You will fall in love with a cat.
111 Maybe.
112 You will fall in love with a stranger in a self-service laundry.
113 Absolutely not.
120 An unexpected check will come by mail.
121 You will trip over a beer can.
122 Not more than $1,000.
123 You will visit an out-of-town friend.
130 You will fall in love with a new car.
131 Positively yes.
132 You will fall in love with a real estate agent.
133 Foolish question.
200 You will dream you are a bird.
201 You never get in brawls.
202 A dream will wake you in the middle of the night.
203 You will have a falling-out with an old friend.
210 You will find a lost key.
211 No injury of any sort for the next seven days, but be careful on the eighth.
212 You will find something unpleasant in your bed.
213 Watch out for a punch on your nose.
220 You will dream of coconut pie.
221 Avoid arguments on a bus.
222 You will dream about a flying saucer.
223 Be careful not to antagonize anyone named Harvey.
230 You will find this trick puzzling.
231 It is a dangerous week to stand on stepladders.
232 You will find the news tomorrow disturbing.
233 Climbing stairways can be dangerous.
300 Yes, lots of money.
301 You will not leave your neighborhood all week.
302 On the contrary, you will lose some money.
303 You will take a marvelous trip in your imagination.
310 You will not fall in love with anyone for a change.
311 You can answer that as well as I can.
312 You will fall for someone in show business.
313 Whom do you think you are kidding?
320 Yes, but most of it will go for taxes.
321 Yes, but you will not enjoy the trip.
322 Some, but you will spend it immediately.
323 You will go on a long trip by plane.
330 You will fall in love twice.
331 I don't know.
332 You will fall out of love.
333 You should be ashamed to ask such a question.
```

you will have a three-digit number that gives the answer to your question.

Larger sets of cards can be designed for answering a larger number of questions. The number of cards must be one less than a power of 2. In 1980 Karl Fulves published *Bob Hummer's Collected Secrets*, a compilation of all known Hummer tricks. This gold mine of ideas for mathematical magic is available from Fulves at Box 433, Teaneck, NJ 07666. Page 77 of the book describes a set of 15 fortunetelling cards, each with four features that may or may not match, to be used with a fortunetelling book (not provided!) of $8^4 = 4,096$ answers. I leave it to readers to puzzle out why the answers are always appropriate.

Having opened with an anonymous limerick about congruences, I shall close with one by John McClellan, an artist living in Woodstock, NY, whose work reflects a lifelong interest in recreational mathematics and wordplay:

> A lady of 80 named Gertie
> Had a boyfriend of 60 named Bertie.
>   She told him emphatically
>   That viewed mathematically
> By modulo 50 she's 30.

Addendum

If any reader is puzzled by "Puthoffa Targu," my suggested precognitive mantra, it is a play on the names of Harold Puthoff and Russell Targ. They were two scientists at Stanford Research International who "confirmed" the psychic powers of Uri Geller. They have since left SRI to go their separate ways, though both remain firmly convinced of the reality of ESP, PK, and precognition.

Phil Goldstein, a magician whose stage name is Max Maven, marketed in 1981, under the name "Mixed Emotions," a clever version of Hummer's witch cards. It comes with seven cards and a booklet of 333 answers titled *The Book of Sins*.

Among hundreds of mathematical card tricks based on congruences, one of the most beautiful involves five cards that are torn in half and apparently randomized by a dealing process. You will find it described as Riddle 5 in my *Riddles of the Sphinx* (Mathematical Association of America, 1987). Magicians have devised numerous variations.

When I said that April 19 was the most frequent date of Easter, I was relying on reported results for short spans of time. Several readers used computers to check time spans longer than Thomas O'Beirne's span of 1900 to 2099. They found that March 31, April 12, and April 15 tied for the most likely date. Thomas L. Lincoln conjectured that with still longer intervals, extending the present rules for Easter far into the future, April 19 would eventually become the commonest date. His conjecture was confirmed by many readers who tested for extremely long future spans. They found April 19 to be the most frequent date, with March 22 the least frequent.

References

NUMBER THEORY AND ITS HISTORY. Oystein Ore. McGraw-Hill Book Company, 1948.

TEN DIVISIONS LEAD TO EASTER. T. H. O'Beirne in *Puzzles and Paradoxes*. Oxford University Press, 1965.

A NEW LOOK AT FUNCTIONS IN MODULAR ARITHMETIC. Marion H. Bird in *The Mathematical Gazette*. Vol. 64, No. 428, pages 78–86; June 1980.

AN APPROACH TO PROBLEM-SOLVING USING EQUIVALENCE CLASSES MODULO n. James E. Schultz and William Burger in *The College Mathematics Journal*. Vol. 15, pages 401–405; November 1984.

CONCRETE MATHEMATICS, Second Edition. Ronald Graham, Donald Knuth, and Oren Patashnik, Chapter 4. Addison-Wesley, 1994.

16

Lavinia Seeks a Room and Other Problems

1 Lavinia Seeks a Room

The line in Figure 96 represents University Avenue in a small college town where Lavinia is a student. The spots labeled A through K are buildings along the avenue in which Lavinia's eleven best friends are living.

Lavinia has been living with her parents in a nearby town, but now she wants to move to University Avenue. She would like a room or an apartment at a location L on the street that minimizes the sum of all its distances from her eleven friends. Assuming that a place is available at

Figure 96

the right location, specify where Lavinia should live and prove it does make the sum of all its distances to the other locations as small as possible.

2   Mirror-Symmetric Solids

On plane figures an axis of symmetry is a straight line that divides the figure into congruent halves that are mirror images of each other. The hearts on playing cards, for example, have one axis of symmetry. So do spades and clubs, but the diamond has two such axes. A square has four axes of symmetry. A regular five-pointed star has five, and a circle has an infinite number. A swastika or a yin–yang symbol has no axis of symmetry.

If a plane figure has at least one axis of symmetry, it is said to be superposable on its mirror image in the following sense. If you view the figure in a vertical mirror, with the edge of the mirror resting on a horizontal plane, you can imagine sliding the figure into the mirror and, if necessary, rotating it on the plane so that it coincides with its mirror reflection. You are not allowed to turn the figure over on the plane because that would require rotating it through a third dimension.

A plane of symmetry is a plane that slices a solid figure into congruent halves, one half a mirror reflection of the other. A coffee cup has a single plane of symmetry. The Great Pyramid of Egypt has four such planes. A cube has nine: three are parallel to a pair of opposite faces and six pass through corresponding diagonals of opposite faces. A cylinder and a sphere each have an infinite number of planes of symmetry.

Think of a solid object bisected by a plane of symmetry. If you place either half against a mirror, with the sliced cross section pressing against the glass, the mirror reflection, together with the bisected half, will restore the shape of the original solid. Any solid with at least one plane of symmetry can be superposed on its mirror image by making, if necessary, a suitable rotation in space.

Discussing this in my book *The Ambidextrous Universe* (Charles Scribner's Sons, 1979), I stated on page 19 that if a three-dimensional object had no plane of symmetry (such as a helix, a Möbius strip, or an overhand knot in a closed loop of rope), it could not be superposed on its mirror image without imagining its making an impossible rotation that "turned it over" through a fourth dimension.

The statement is false! As many readers of the book pointed out, there are solid figures totally lacking a plane of symmetry that nonetheless can, by a suitable rotation in ordinary space, be superposed on their mirror images. In fact, one is so simple that you can fold it in a trice from a square sheet of paper. How is it done?

3   The Damaged Patchwork Quilt

The patchwork quilt in Figure 97, of size 9 by 12, was originally made up of 108 unit squares. Parts of the quilt's center became worn, making it necessary to remove eight squares as indicated.

The problem is as follows. Cut the quilt along the lattice lines into

Figure 97

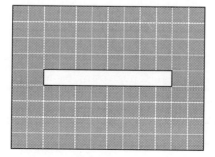

just two parts that can be sewn together to make a 10-by-10 square quilt. The new quilt cannot, of course, have any holes. Either part can be rotated, but neither one can be turned over because the quilt's underside does not match its upper side.

The puzzle is an old one, but the solution is so beautiful and the problem is so little known that I constantly get letters about it from readers who are not aware of its origin. The solution is unique even if it is not required that the cutting be along lattice lines.

4   Acute and Isosceles Triangles

An acute triangle is one with each interior angle less than 90 degrees. What is the smallest number of nonoverlapping acute triangles into which a square can be divided?

I asked myself that question some twenty years ago, and I solved it by showing how to cut a square into eight acute triangles as is indicated in Figure 98, top. Reporting this in a column, reprinted as Chapter 3 of my *New Mathematical Diversions from Scientific American* (Simon and Schuster, 1966), I said: "For days I was convinced that nine was the answer; then suddenly I saw how to reduce it to eight."

Since then I have received many letters from readers who were unable to find a solution with nine acute triangles but who pointed out that solutions are possible for ten or any higher number. The middle illustration in Figure 98 shows how it is done with ten. Note that obtuse triangle ABC is cut into seven acute triangles by a pentagon of five acute triangles. If ABC is now divided into an acute and an obtuse triangle by BD, as is indicated in Figure 97, bottom, we can use the same pentagonal method for cutting the obtuse triangle BCD into seven acute triangles, thereby producing eleven acute triangles for the entire square. A repetition of the procedure will produce 12, 13, 14, . . . acute triangles.

Apparently the hardest dissection to find is the one with nine acute triangles. Nevertheless, it can be done.

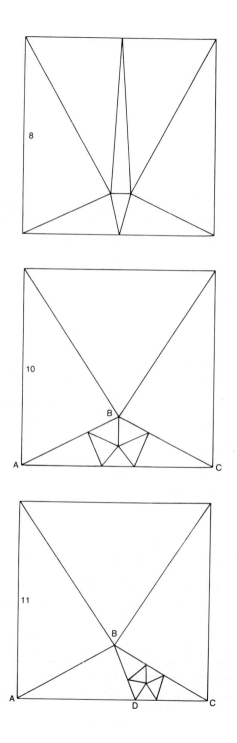

Figure 98

There are many comparable problems about cutting figures into nonoverlapping triangles, of which I shall mention only two. It is easy to divide a square into any even number of triangles of equal area, but can a square be cut into an odd number of such triangles? The surprising answer is no. As far as I know, this was first proved by Paul Monsky in *American Mathematical Monthly* (Vol. 77, No. 2, pages 161–164; February 1970).

Another curious theorem is that any triangle can be cut into n isosceles triangles provided n is greater than 3. A proof by Gali Salvatore appeared in *Crux Mathematicorum* (Vol. 3, No. 5, pages 134–135; May 1977). Another proof, by N. J. Lord, is in *The Mathematical Gazette* (Vol. 66, pages 136–137; June 1982).

The case of the equilateral triangle is of particular interest. It is easy to cut it into four isosceles triangles (all equilateral) or into three isosceles triangles. (Some triangles cannot be cut into three or two isosceles triangles, which is why the theorem requires that n be 4 or more.) Can you cut an equilateral triangle into five isosceles triangles? I shall show how it can be done with none of the five triangles equilateral, with just one equilateral, and with just two equilateral. It is not possible for more than two of the isosceles triangles to be equilateral.

5 Measuring with Yen

This problem was originated by Mitsunobu Matsuyama, a reader in Tokyo. He sent me a supply of Japanese one-yen coins and told me of the following remarkable facts about them that are not well known even in Japan. The one-yen coin is made of pure aluminum, has a radius of exactly one centimeter, and weighs just one gram. Thus a supply of yens can be used with a balance scale for determining the weight of small objects in grams. It also can be used on a plane surface for measuring distances in centimeters.

It is easy to see how one-yen coins can be placed on a line to measure distances in even centimeters (two centimeters, four, six, and so on), but can they also be used to measure odd distances (one, three, five, and so on)? Show how a supply of one-yen coins can be used for measuring all integral distances in centimeters along a line.

6 A New Map-Coloring Game

This problem comes to me from its originator, Steven J. Brams, a political scientist at New York University. He is the author of *Game Theory and Politics* (1975), *Paradoxes in Politics* (1976), and *The Presidential Election Game* (1978). His *Biblical Games* (The MIT Press, 1980) is a surprising application of game theory to Old Testament episodes of a game-like nature in which one of the players is assumed to be an omniscient deity. This was followed by *Superior Beings: If They Exist, How Would We Know?* (1983), *Rational Politics* (1985), *Superpower Games* (1985), *Negotiating Games* (1990), and *Theory of Moves* (1994). His latest book, written with Alan D. Taylor, is *Fair Division* (1996), an analysis of cake cutting and other fair division problems.

Suppose we have a finite, connected map on a plane and a supply of n crayons of different colors. The first player, the minimizer, selects any crayon and colors any region on the map. The second player, the maximizer, then colors any other region, using any of the n colors. Players continue in this way, alternately coloring a region with any of the n colors but always obeying the rule that no two regions of the same color can share any portion of a common border. Like colors may, of course, touch at points.

The minimizer tries to obviate the need for an $n + 1$ color to complete the map. The maximizer tries to force the use of it. The maximizer wins if either player is unable to play, using one of the n crayons, before the map is fully colored. If the entire map is colored with n colors, the minimizer wins.

The subtle and difficult problem is: What is the smallest value of n such that when the game is played on any map, the minimizer can, if both players play optimally, always win?

To make the problem clearer, consider the simple map in Figure 99. It proves that n is at least 5. Of course, if no game is played, the map can be easily colored with four colors, as indeed any map on the plane can be. (This is the famous four-color theorem, now known to be true.) But if the map is used for playing Brams's game, and only four colors are available, the maximizer can always force the minimizer to use a fifth color. If five colors are available, the minimizer can always win.

Brams conjectures that the minimum value of n is 6. A map has been found on which five colors allow the maximizer always to win. Can you construct such a map and give the maximizer's winning strategy? Remember, the minimizer goes first, and neither player is under any compulsion to introduce a new color at any turn if he can legally play a color that has already been used.

Figure 99

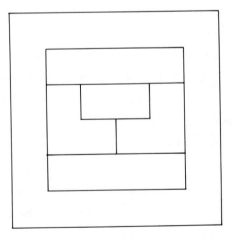

7   Whim

In his Pulitzer-prize book *Gödel, Escher, Bach*, Douglas R. Hofstadter introduces the concept of self-modifying games. These are games in which on his turn a player is allowed, instead of making a legal move, to announce a new rule that modifies the game. The new rule is called a metarule. A rule that modifies a metarule is a metametarule, and so on. Hofstadter gives some chess examples. Instead of moving, a player might announce that henceforth a certain square may never be occupied, or that all knights must move in a slightly different manner, or any other metarule that is on a list of allowed alterations of the game.

The basic idea is not entirely new; before 1970 John Horton Conway proposed a whimsical self-modifying variation of nim that he called whim. Nim is a two-person game played with counters that are arranged in an arbitrary number of piles, with an arbitrary number of counters in each pile. Players take turns removing one or more counters from any one pile. In normal nim the person taking the last counter wins; in misère, or reverse nim, the person taking the last counter loses. The strategy for perfect play has long been known. You will find it in the chapter on nim in my *Scientific American Book of Mathematical Puzzles & Diversions* (Simon and Schuster, 1959).

Whim begins without any decision on whether the game is normal or misère. At any time in the game, however, either player may, instead of making a move, announce whether the game is normal or misère. This "whim move" is made only once; from then on the game's form is frozen. It is well known that in nim the strategy is the same for both forms of the game until near the end, so that you may be tempted to suppose whim strategy is easily analyzed. Try playing a few games and you will find it is not as simple as it seems!

Suppose you are the first to play in a game with many piles and many counters in each pile, and the position is a nim loss for you. You should at once make the whim move because it leaves the position unchanged and you become the winning player. Suppose, however, you are first to play and the position is a win for you. You dare not make a winning move because this allows your opponent to invoke the whim and leave you with a losing position. Hence you must make a move that would lose in ordinary nim. For the same reason your opponent must follow with a losing move. Of course, if one player fails to make a losing move, the other wins by invoking the whim.

As the game nears the end, reaching the point where the winning strategy diverges for normal or for reverse nim, it may be necessary to invoke the whim in order to win. How is this determined? And how can one decide at the start of a game who has the win when both sides play as well as possible? Conway's strategy is easy to remember but, as he once remarked, hard to guess even by someone well versed in nim theory.

Answers

1   Lavinia Seeks a Room

Consider the two outermost spots, A and K. A spot L on the line anywhere between A and K (inclusive) will have the same sum of its distances to A and K. Clearly this is smaller than the sum if L is not between A and K. Now consider B and J, the next pair of spots as you move inward. As before, to minimize the sum of L's distances from B and J, L must be between B and J. Since L is then also between A and K, its position will minimize the sum of its distances to A, K, B, and J.

Continue in this way, taking the spots by pairs as you move inward to new nested intervals along the line. The last pair of spots is E and G. Between them is the single spot F. Any spot between E and G will minimize the distances to all spots except F. Obviously if you want also to minimize the distance to F, the spot must be exactly at F. In brief, Lavinia should move into the same building where her friend Frank lives.

To generalize, for any even number of spots on a straight line, any spot between the two middle spots will have a minimal sum of distances to all spots. For any odd number of spots the center spot is the desired location. The problem appeared in "No Calculus, Please," by J. H. Butchart and Leo Moser (*Scripta Mathematica,* Vol. 18, Nos. 3-4, pages 221-236; September-December 1952).

2   Mirror-Symmetric Solids

Figure 100 shows how to fold a square of paper into a shape that has no plane of symmetry yet can be superposed on its mirror reflection. The figure is said to have "mirror-rotation symmetry," a type of symmetry of great importance in crystallography. I took this example from page 42 of *Symmetry in Science and Art,* by A. V. Shubnikov and V. A. Koptsik (Plenum Press, 1974).

3   The Damaged Patchwork Quilt

Figure 101 shows how the damaged patchwork quilt is cut into two parts that can be sewn together to make a square quilt without a hole. The problem is No. 215 in Henry Ernest Dudeney's *Puzzles and Curious Problems* (Thomas Nelson and Sons, Ltd., 1931).

Figure 100

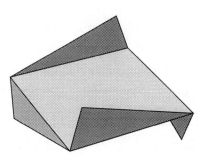

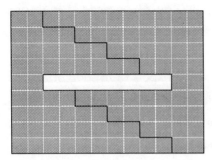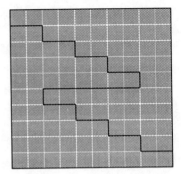

Figure 101

4   Acute and Isosceles Triangles

Figure 102 shows how to cut a square into nine acute triangles. The solution is unique. If triangulation is taken in the topological sense, so that a vertex is not allowed to be on the side of a triangle, then there is no solution with nine triangles, although there is a solution for eight triangles, for 10 and for all higher numbers. This curious result has been proved in an unpublished paper by Charles Cassidy and Graham Lord of Laval University in Quebec.

Figure 103 shows four ways to cut an equilateral triangle into five

Figure 102

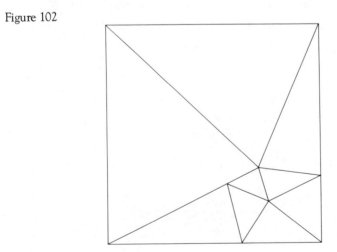

isosceles triangles. The first pattern has no equilateral triangle among the five, the second and third patterns both have one equilateral triangle and the fourth pattern has two equilateral triangles. The four patterns, devised by Robert S. Johnson, appear in *Crux Mathematicorum* (Vol. 4, No. 2, page 53; February 1978). A proof by Harry L. Nelson that there cannot be more than two equilateral triangles is in the same volume of the journal (No. 4, pages 102–104; April 1978).

The first three patterns of Figure 101 are not unique. Many readers sent alternate solutions. The largest number, 13, came from Roberto Teodoro Garrido, a civil engineer in Buenos Aires.

Figure 103

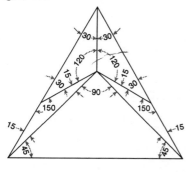
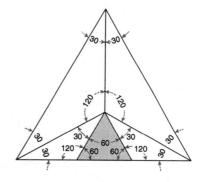
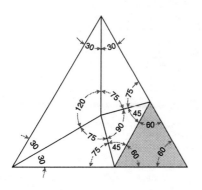
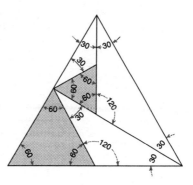

5   Measuring with Yen

Integral distances in centimeters along a line can be measured by one-yen Japanese coins, each with a radius of one centimeter, in the manner shown in Figure 104.

6   A New Map-Coloring Game

The map shown in Figure 105 found by Lloyd Shapley, a mathematician at the Rand Corporation, proves that when five colors are used in playing Steven J. Brams's map-coloring game, there is a map on which the maximizer can always win.

The map is a projection of the skeleton of a dodecahedron, with an outside region (A) that represents the "back" face of the solid. The maximizer's strategy is always to play on the face of the dodecahedron opposite the face where the opponent last played, using the same color. (In the illustration regions representing pairs of opposite faces are given

Figure 104

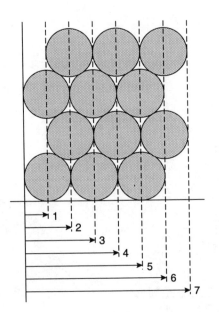

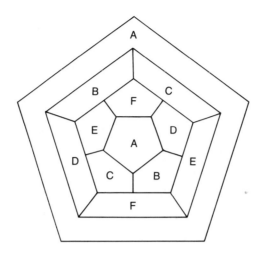

Figure 105

the same letter.) As you can readily see, this strategy eliminates successive colors from further use, forcing the game to the point where the minimizer cannot play without using a sixth color.

Is there a map on which the maximizer can force the use of a seventh color when the opponent plays optimally? This remains an unsolved problem.

7   Whim

John Horton Conway's strategy for his game of whim is as follows. Treat the whim move as if it were another pile consisting of one counter if there is a pile of four or more counters, and as if it were another pile consisting of two counters if there is no pile of four or more. Until someone makes the whim move the invisible whim pile remains. Making the whim move removes the whim pile. A whim position is a winning position for the player who would win in ordinary nim if the whim pile were actually there. The winning strategy, therefore, is simply to imagine the whim is present until a player removes it and to play the strategy of ordinary nim. If a move changes a position from one in which at least one pile has four or more counters to one in which there is no such pile, the whim pile acquires its second counter after the move, not before.

ADDENDUM

Thomas Szirtes made the following comment on the problem about Lavinia's room: "The solution is interesting because it is contrary to intuition. According to the rules, the location of the minimum-sum distance is independent of the relative or even the absolute distances among the points. One would feel, for example, that if the rightmost point K receded 100 miles, it would somehow 'pull' the minimum sum distance point to the right. But this is not the case. In fact, all points to the right of F could recede to infinity and all points to the left of F could be infinitely close to F, and the minimum sum distance point still would be at F!"

I showed how a square sheet of paper can be folded obliquely along four sides to produce a shape that lacks any plane of symmetry yet is superposable on its mirror image. Paul Schwink of Carlisle, Iowa, and Piet Hein of Copenhagen pointed out that the solution is unnecessarily complicated. The same result can be obtained by folding just one pair of opposite sides, one up, one down.

The problem of the mutilated quilt leads to a variety of generalizations. In the problem given, if the rectangle is chessboard colored, the 10-by-10 square will not preserve such coloring. I spent several days exploring squares of side n, with an associated rectangle of $(n - 1) \times (n + 2)$ sides, with a hole of $1 \times (n - 2)$ sides parallel to the rectangle's long side and as centered as possible. Assuming the rectangle to be chessboard colored, reader Eric Stott asked, Into how few pieces can it be sliced along lattice lines to form a properly colored square?

Consider the rectangle shown in Figure 106. I found it could be sliced into as few as three pieces that will form a standard chessboard as shown at the top of Figure 107. The generalization to squares of even-numbered sides is obvious. If the square's side is odd, one must distinguish between $n = 1$

Figure 106

Figure 107

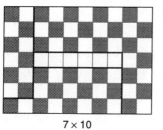

7 × 10   8 × 8

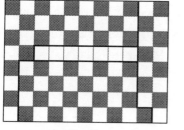
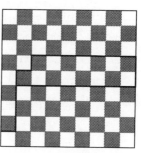

8 × 11   9 × 9

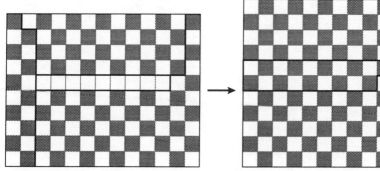

10 × 13   11 × 11

(mod 4) and $n = 3$ (mod 4). Examples of each are shown in the middle and bottom illustrations. Again it is easy to see how they generalize.

Two-piece dissections are impossible; four-piece dissections are plentiful and easy to find. In all such solutions with three pieces I believe it is impossible to avoid mirror reflecting the asymmetric piece.

The task of minimizing the number of acute triangles into which a square can be cut suggests companion problems for right triangles, obtuse triangles, and scalene triangles. For right triangles the trivial answer is two; for scalene triangles the easy answer is three. Figure 108 shows how a square can be cut into six obtuse triangles. I believe this to be minimal.

Into how many equilateral triangles can an equilateral triangle be cut? Two and three are impossible, four is obvious, five is impossible, and any number above five is possible.

Steven J. Brams's map-coloring game led Robert High, a mathematician with Informatics, Inc., of New York City, to some surprising results. Call the first player Min (who tries to minimize colors) and the second player Max (who tries to maximize colors). I had given a map of six regions on which Max can force five colors. High found that a projection of the cube gives a simpler six-region map for forcing five colors. Max merely places at each turn a new color on the face opposite Min's last play.

Figure 108

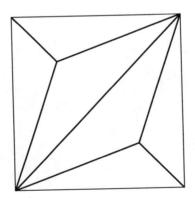

The map forcing six colors, which I gave, was a projection of the skeleton of a dodecahedron. High found that if four corners of a cube are replaced by triangular faces, as shown in Figure 109, a projection gives a map of only ten regions on which Max can force six colors. The strategy is less elegant than the one on the dodecahedral 12-region map and is a bit too involved to give here.

The biggest surprise was High's discovery of a 20-region map on which Max can force seven colors! Imagine each corner of a tetrahedron replaced by a triangular face, then each of the 12 new corners is replaced by a triangular face. Figure 110 shows a planar projection of the resulting polyhedron's skeleton. Here the strategy is more complex than the one for High's 10-region map, but he sent a game tree that proves the case. High conjectures that no planar map allows Max to force more than seven colors, but this remains unproved. It is even possible that there is no minimum upper bound.

Figure 109

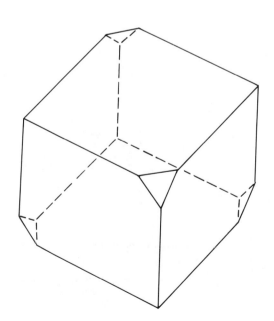

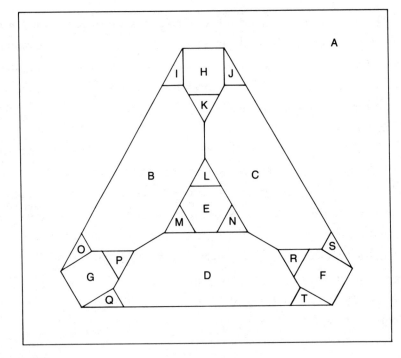

Figure 110

17

The
Symmetry
Creations
of
Scott Kim

S cott Kim's *Inversions*, published in 1981 by Byte Publications, is one of the most astonishing and delightful books ever printed. Over the years Kim has developed the magical ability to take just about any word or short phrase and letter it in such a way that it exhibits some kind of striking geometrical symmetry. Consider Kim's lettering of my name in Figure 111. Turn it upside down and presto! It remains exactly the same.

Students of curious wordplay have long recognized that short words

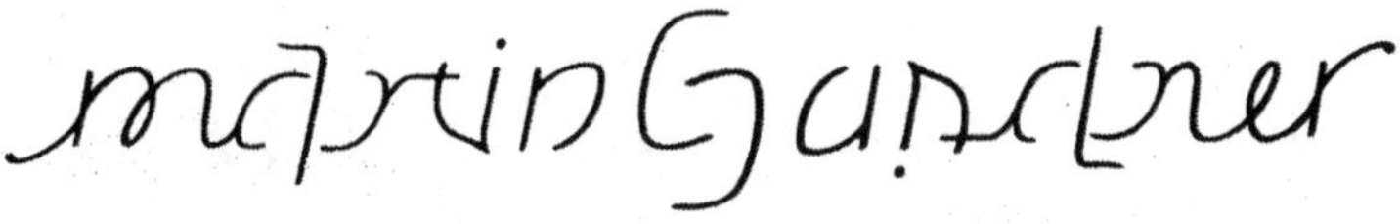

martin Gardner

Figure 111

can be formed to display various types of geometrical symmetry. On the Rue Mozart in Paris a clothing shop called "New Man" has a large sign lettered "NeW MaN" with the e and the a identical except for their orientation. As a result the entire sign has upside-down symmetry. The names VISTA (the magazine of the United Nations Association), ZOONOOZ (the magazine of the San Diego zoo) and NISSIN (a Japanese manufacturer of camera flash equipment) are all cleverly designed so that they have upside-down symmetry.

BOO HOO, DIOXIDE, EXCEEDED and DICK COHEN DIED 10 DEC 1883 all have mirror symmetry about a horizontal axis. If you hold them upside down in front of a mirror, they appear unchanged. One day in a supermarket my sister was puzzled by the name on a box of crackers, "spep oop," until she realized that a box of "doo dads" was on the shelf upside down. Wallace Lee, a magician in North Carolina, liked to amuse friends by asking if they had ever eaten any "ittaybeds," a word he printed on a piece of paper like this:

Ittaybeds

After everyone said no, he would add:

"Of course, they taste much better upside down."

Many short words in conventional typefaces turn into other words when they are inverted. MOM turns into WOW and "up" becomes the abbreviation "dn." SWIMS remains the same. Other words have mirror symmetry about a vertical axis, such as "bid" (and "pig" if the g is drawn as a mirror image of the p). Here is an amusing way to write "minimum" so that it is the same when it is rotated 180 degrees:

It is Kim who has carried this curious art of symmetrical calligraphy to heights not previously known to be possible. By ingeniously distorting letters, yet never so violently that one cannot recognize a word or phrase, Kim has produced incredibly fantastic patterns. His book is a collection of such wonders, interspersed with provocative observations on the nature of symmetry, its philosophical aspects, and its embodiment in art and music as well as in wordplay.

Kim is no stranger to my *Scientific American* columns. He is a young man of Korean descent, born in the U.S., who in 1981 was doing graduate work in computer science at Stanford University. He was in his teens when he began to create highly original problems in recreational mathematics. Some that have been published in *Scientific American* include his "lost-king tours" (April 1977), the problem of placing chess knights on the corners of a hypercube (February 1978), his solution to "boxing a box" (February 1979), and his beautifully symmetrical "*m*-pire map" given here in Chapter 6. In addition to a remarkable ability to think geometrically (not only in two and three dimensions but also in four-space and higher spaces) Kim is a classical pianist who for years could not decide between pursuing studies in mathematics or in music. At the moment he is intensely interested in the use of computers for designing typefaces, a field pioneered by his friend and mentor at Stanford, the computer scientist Donald E. Knuth.

For several years Kim's talent for lettering words to give them unexpected symmetries was confined to amusing friends and designing family Christmas cards. He would meet a stranger at a party, learn his or her name, then vanish for a little while and return with the name neatly drawn so that it would be the same upside down. His 1977

Christmas card, with upside-down symmetry, is shown in Figure 112. (Lester and Pearl are his father and mother; Grant and Gail are his brother and sister.) The following year he found a way to make "Merry Christmas, 1978," mirror-symmetrical about a horizontal axis, and in 1979 he made the mirror axis vertical. (See Figures 113 and 114.)

For a wedding anniversary of his parents Kim designed a cake with chocolate and vanilla frosting in the pattern shown in Figure 115. ("Lester" is in black, "Pearl" is upside down in white.) This is Kim's "figure and ground" technique. You will find another example of it in *Gödel*,

Figure 112

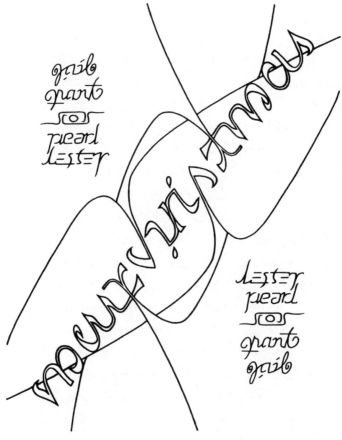

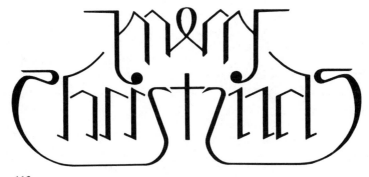

Figure 113

Figure 114

Figure 115

Escher, Bach: An Eternal Golden Braid, the Pulitzer-prize-winning book by Kim's good friend Douglas R. Hofstadter. Speaking of Kurt Gödel, J. S. Bach and M. C. Escher, Figure 116 shows how Kim has given each name a lovely mirror symmetry. In Figure 117 Kim has lettered the entire alphabet in such a way that the total pattern has left–right symmetry.

Kim's magic calligraphy came to the attention of Scot Morris, an editor at *Omni*. Morris devoted a page of his popular column on games

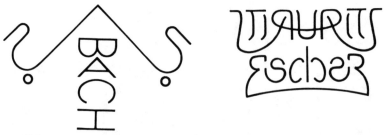

Figure 116

to Kim's work in *Omni*'s September 1979 issue, and he announced a reader's contest for similar patterns. Kim was hired to judge the thousands of entries that came in. You will find the beautiful prizewinners

Figure 117

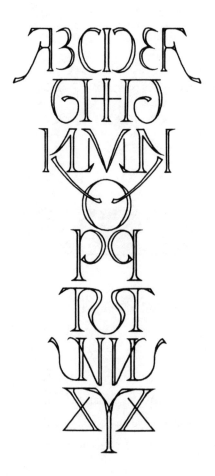

in *Omni*'s April 1980 issue and close runners-up in Morris' columns for May and November of the same year.

All the patterns in Kim's book are his own. A small selection of a few more is given in Figure 118 to convey some notion of the amazing variety of visual tricks Kim has up his sleeve.

I turn now to two unusual mathematical problems originated by Kim, both of which are still only partly solved. In 1975, when Kim was

Figure 118

in high school, he thought of the following generalization of the old problem of placing eight queens on a chessboard so that no queen attacks another. Let us ask, said Kim, for the maximum number of queens that can be put on the board so that each queen attacks exactly n other queens. As in chess, we assume that a queen cannot attack through another queen.

When n is 0, we have the classic problem. Kim was able to prove that when n is 1, 10 queens is the maximum number. (A proof is in *Journal of Recreational Mathematics*, Vol. 13, No. 1, page 61; 1980–81.) A pleasing solution is shown in Figure 119, at the top. The middle illustration shows a maximal solution of 14 queens when n is 2, a pattern Kim described in a letter as being "so horribly asymmetric that it has no right to exist." There are only conjectures for the maximum when n is 3 or 4. Kim's best result of 16 queens for $n = 3$ has the ridiculously simple solution shown in Figure 119, bottom, but there is no known proof that 16 is maximum. For $n = 4$ Kim's best result is 20 queens. Can you place 20 queens on a chessboard so each queen attacks exactly four other queens?

The problem can of course be generalized to finite boards of any size, but Kim has a simple proof based on graph theory that on no finite board, however large, can n have a value greater than 4. For $n = 1$ Kim has shown that the maximum number of queens cannot exceed the largest integer less than or equal to $4k/3$, where k is the number of squares along an edge of the board. For $n = 2$ he has a more difficult proof that the maximum number of queens cannot exceed $2k - 2$, and that this maximum is obtainable on all even-order boards.

Kim's problem concerning polycube snakes has not previously been published, and he and I would welcome any light that readers can throw on it. First we must define a snake. It is a single connected chain of identical unit cubes joined at their faces in such a way that each cube (except for a cube at the end of a chain) is attached face to face to exactly

Figure 119

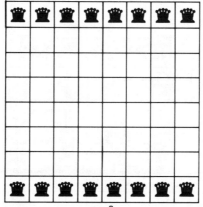

n = 1

n = 2

n = 3

two other cubes. The snake may twist in any possible direction, provided no internal cube abuts the face of any cube other than its two immediate neighbors. The snake may, however, twist so that any number of its cubes touch along edges or at corners. A polycube snake may be finite in length, having two end cubes that are each fastened to only one cube, or it may be finite and closed so that it has no ends. A snake may also have just one end and be infinite in length, or it may be infinite and endless in both directions.

We now ask a deceptively simple question. What is the smallest number of snakes needed to fill all space? We can put it another way. Imagine space to be completely packed with an infinite number of unit cubes. What is the smallest number of snakes into which it can be dissected by cutting along the planes that define the cubes?

If we consider the two-dimensional analogue of the problem (snakes made of unit squares), it is easy to see that the answer is two. We simply intertwine two spirals of infinite one-ended flat snakes, one gray, one white, as in Figure 120.

The question of how to fill three-dimensional space with polycube snakes is not so easily answered. Kim has found a way of twisting four

Figure 120

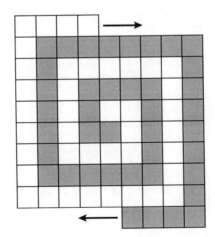

infinitely long one-ended snakes (it is convenient to think of them as being each a different color) into a structure of interlocked helical shapes that fill all space. The method is too complicated to explain in a limited space; you will have to take my word that it can be done.

Can it be done with three snakes? Not only is this an unanswered question but also Kim has been unable to prove that it cannot be done with two! "A solution with only two snakes," he wrote in a letter, "would constitute a sort of infinite three-dimensional yin–yang symbol: the negative space left by one snake would be the other snake. It is the beauty of such an entwining, and the possibility of building a model large enough to crawl through, that keeps me searching for a solution."

The problem can of course be generalized to snakes made of unit cubes in any number of dimensions. Kim has conjectured that in a space of n dimensions the minimum number of snakes that completely fill it is $2(n - 1)$, but the guess is still a shaky one.

A few years ago I had the pleasure of explaining the polycube-snake problem to John Horton Conway, the Cambridge mathematician. When I concluded by saying Kim had not yet shown that two snakes could not tile three-dimensional space, Conway instantly said, "But it's obvious that—" He checked himself in mid-sentence, stared into three-space for a minute or two, then exclaimed, "It's *not* obvious!"

I have no idea what passed through Conway's mind. I can only say that if the impossibility of filling three-space with two snakes is not obvious to Conway or to Kim, it probably is not obvious to anyone else.

Answers

Readers were asked to place 20 queens on a chessboard so that each queen attacks exactly four others. A solution is given in Figure 121.

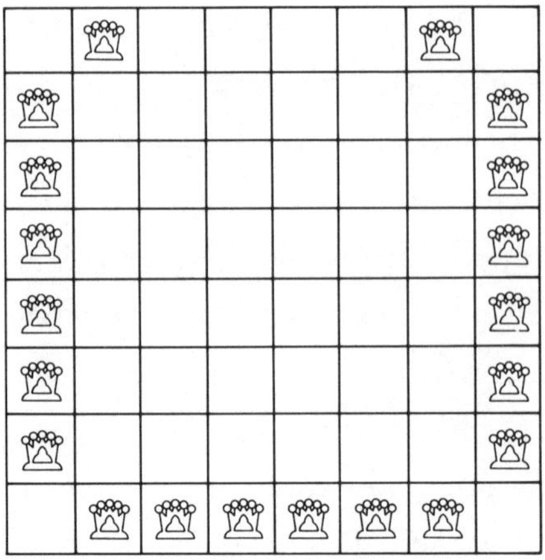

Figure 121

ADDENDUM

Dozens of readers sent examples of printed words and even sentences that are unreversed in mirrors, or which change to other words. (See Figure 122 for an example.) Several readers noticed that TOYOTA, written vertically, is unaltered by a mirror. I discovered that the following garbled sentence:

Figure 122

MOM

TOP

OTTO

A

GOT

reads properly when you see it in a mirror.

David Morice published this two-stanza "poem" in *Wordways* (November 1987, page 235).

DICK HID

CODEBOOK +

DOBIE KICKED

HOBO–OH HECK–I DECIDED

I EXCEEDED ID–I BOXED

HICK–ODD DODO–EH KID

DEBBIE CHIDED–HOCK CHECKBOOK

ED–BOB BEDDED CHOICE CHICK

HO HO–HE ECHOED–OH OH

DOBIE ICED HOODED IBEX

I COOKED OXHIDE COD

EDIE HEEDED COOKBOOK +

ED

DECKED

BOB

To read the second stanza, hold the poem upside down in front of a mirror.

Donald Knuth, Ronald Graham, and Oren Patashnik, in their marvelous book *Concrete Mathematics* (the word is a blend of *Continuous* and *Discrete* mathematics), published by Addison-Wesley in 1989, introduce their readers to the "umop-apisdn" function. Rotate the word 180 degrees to see what it means in English.

One conjecture about the origin of the expression "Mind your *p*s and qs" is that printers often confused the two letters when they were in lower-case. A more plausible theory is that British tavern owners had to mind their pints and quarts.

In his autobiography *Arrow in the Blue*, Arthur Koestler recalls meeting many science cranks when he was a science editor in Berlin. One was a man who had invented a new alphabet. Each letter had fourfold rotational symmetry. This, he proclaimed, made it possible for four people, seated on the four sides of a table, to simultaneously read a book or newspaper at the table's center.

Have you heard about the dyslexic atheist who didn't believe in dog? Or D.A.M.N., an organization of National Mothers Against Dyslexia?

I could easily write another chapter about the amazing Scott Kim. He received his Ph.D. in *Computers and Graphic Design*, at Stanford University, working under Donald Knuth. At a curious gathering of mathematicians, puzzle buffs, and magicians, in Atlanta in 1995, Kim demonstrated how your fingers can model the skeleton of a tetrahedron and a cube, and how they can form a trefoil knot of either handedness. He also played an endless octave on a piano, each chord rising up the scale yet never going out of hearing range, and proved he could whistle one tune and hum another at the same time. During the Atlanta gathering, he and his friends Karl Schaffer and Erik Stern of the Dr. Schaffer & Mr. Stern Dance Ensemble presented a dance performance titled "Dances for the Mind's Eye." Choreographed by the three performers, the performance was based throughout on mathematical symmetries.

Among books illustrated by Kim are my *Aha! Gotcha* (W. H. Freeman) and Ilan Cardi's *Illustrated Computational Recreations in Mathematica* (Addison-Wesley). Together with Ms. Robin Samelson, Kim produced *Letterform and Illusion*, a computer disk with an accompanying 48-page book of programs designed for use with Claris's MacPaint. In 1994 Random House published Kim's *Puzzle Workout*, a collection of 42 brilliant puzzles reprinted from his puzzle column in *New Media Magazine*. It is the only book of puzzles known to me in which every single puzzle is totally original with the author.

Scott Kim's queens problem brought many letters from readers who sent variant solutions for $n = 2$, 3, and 4 on the standard chessboard, as well as proofs for maximum results, and unusual ways to vary the problem. The most surprising letters came from Jeffrey Spencer, Kjell Rosquist, and William Rex Marshall. Spencer and Rosquist, writing in 1981, each independently bettered by one Kim's 20-queen solution for $n = 4$ on the chessboard. Figure 123 shows how each placed 21 queens. It is not unique. Writing in 1989 from Dunedin, New Zealand, Marshall sent 36 other solutions!

Marshall also went two better than Kim's chessboard pattern for $n = 3$. He sent nine ways that 18 queens can each attack three others on the chessboard. The solution shown in Figure 124 is of special interest because only three queens are not on the perimeter. Marshall found a simple pigeonhole proof that for the order-8 board, $n = 4$, 21 queens is indeed maximum. His similar proof shows 18 maximum for $n = 3$. More generally, he showed that for $n = 4$, with k the order of the board, the maximum is $3k - 3$ for k greater than 5. When $n = 3$, Marshall proved that the maximum number of queens is the largest even number less than or equal to $(12k - 4)/5$. For $n = 2$, he found that Kim's formula of $2k - 2$ applies to all boards larger than order 2, not just to boards of even order.

Figure 123

Figure 124

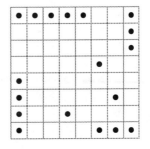

Perhaps it is worth noting that when $n = 4$, no queen can occupy a corner cell because there is no way it can attack more than three other queens. Dean Hoffman sent a simple proof that n cannot exceed 4. Consider the topmost queen in the leftmost occupied row. At the most it can attack four other queens.

In 1991 Peter Hayes sent a letter from Melbourne, Australia, in which he independently obtained the same results, including all proofs, as those obtained by William Marshall. They were published in a paper titled "A Problem of Chess Queens," in the *Journal of Recreational Mathematics*, Vol. 24, pages 264–271; No. 4, 1992.

In 1996 I received a second letter from William Marshall. He sent me the results of his computer program which provided complete solutions of Kim's chess problem for k (the order of the board) = 1 through 9, and n (number of attacked queens) = 1 through 4. A chart extending these results to $k = 10$ is shown in Figure 125.

Note that in four cases there are unique solutions. These are shown in Figure 126. Figure 127 shows a second solution for $n = 2$, $k = 8$, and Figure 128 is an elegant solution for $n = 3$, $k = 9$ found among the 755 patterns produced by Marshall's program.

Dr. Koh Chor Jin, a physicist at the National University of Singapore, sent a clever proof that given a finite volume of space it is possible to cover it with two of Kim's cube-connected snakes. However, as Kim pointed out, Jin's construction does not approach all of space as a limit as the volume of space

Figure 125

k	$n = 1$	$n = 2$	$n = 3$	$n = 4$
3	0	4	2	0
4	5	2	4	0
5	0	1	31	0
6	2	1	304 (307)	1
7	138 (149)	5	2	3
8	47 (49)	2	9	40
9	1	15	755	615
10	12,490 (12,897)	3	39,302	16,573

increases. Each time you wish to enlarge his construction, it has to be modified. Kim is convinced that tiling all of space with two snakes is impossible, but for three snakes the question remains open.

Figure 126

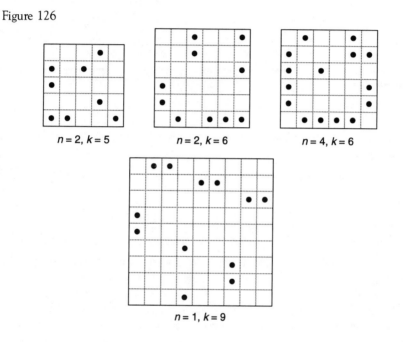

n = 2, k = 5   n = 2, k = 6   n = 4, k = 6

n = 1, k = 9

Figure 127   Figure 128

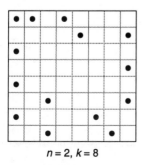   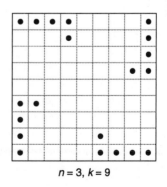

n = 2, k = 8

n = 3, k = 9

References

SYMMETRY. Hermann Weyl. Princeton University Press, 1952.

ROTATIONS AND REFLECTIONS. Martin Gardner in *The Unexpected Hanging and Other Mathematical Diversions*. Simon and Schuster, 1969.

SYMMETRY IN SCIENCE AND ART. A. V. Schubnikov and V. A. Koptsik. Plenum Press, 1974.

INVERSIONS. Scott Kim. Byte, 1981, Key Curriculum Press, 1996.

AMBIGRAMMI. Douglas Hofstadter. Foreword by Scott Kim. Florence, Italy: Hopefulmonster, 1987. An English edition, no date, has been printed by Hofstadter.

WORDPLAY: AMBIGRAMS AND REFLECTIONS ON THE ART OF AMBIGRAMS. John Langdon. Foreword by Martin Gardner. Harcourt Brace Jovanovich, 1992.

FEARFUL SYMMETRY. Ian Stewart and Martin Golubitsky. Blackwell, 1992.

SYMMETRY: A UNIFYING CONCEPT. István and Magdolna Hargittai. Shelter Publications, 1994.

18

Parabolas

Mathematicians are constantly constructing and exploring the properties of abstract objects only because they find them beautiful and interesting. Later, sometimes centuries later, the objects may turn out to be enormously useful when they are applied to the physical world. There are no more elegant examples of this than the work done in ancient Greece on the four conic-section curves. Earlier articles of mine have dealt with three of them: circles, ellipses, and hyperbolas. This time we take a look at parabolas.

If a right circular cone is sliced by a plane parallel to its base, the cross section is a circle. Tip the plane ever so slightly and the section becomes an ellipse, the locus of all points with distances from two fixed points (foci) that have a constant sum. Think of the circle as an ellipse with foci that have merged to become the center of the circle. As the cutting plane tips at progressively steeper angles, the two foci move farther apart and the ellipses become progressively more "eccentric." When the plane is exactly parallel to the side of the cone, the cross section is a parabola. It is a limit curve, like the circle, only now one focus has vanished by moving off to infinity. It is an ellipse, as Henri Fabre once put it, that "seeks in vain for its second, lost center."

As you follow the parabola's arms toward infinity, they get progressively closer to being parallel without ever making it except at infinity. Here is how Johannes Kepler put it in a discussion of conic sections:

> Because of its intermediate nature the parabola occupies a middle position [between the ellipse and the hyperbola]. As it is produced it does not spread out its arms like the hyperbola but contracts them and brings them nearer to parallel, always encompassing more, yet always striving for less—whereas the hyperbola, the more it encompasses, the more it tries to obtain.

A parabola is the locus of all points in a plane whose distance from a fixed line (the directrix) is equal to its distance from a fixed point (focus) not on the line. Figure 129 shows the traditional way of graphing a parabola so that its Cartesian-coordinate equation is extremely simple. Note that the axis of the parabola passes through the focus at right angles to the directrix, and that the tip of the curve, called the vertex, is at the 0,0 point of origin. The chord passing through the focus, perpendicular to the axis is the parabola's latus rectum, or focal width. Let a be the distance from the focus to the vertex. Obviously a

is also the distance from the vertex to the directrix. It is not hard to prove that the latus rectum must be 4a. The parabola can now be described as the locus of all points on the Cartesian plane given by the parabola's type equation: $y^2 = 4ax$. If $y^2 = x^2$, the parabola's vertex is at 0,0. More generally, any quadratic equation of the form $x = ay^2 + by + c$, where a is not zero, graphs as a parabola, although not necessarily a parabola positioned like the typical one shown in the illustration.

A surprising property of the parabola is that all parabolas have the same shape. To be sure, pieces of parabolas, like the two shown in Figure 130, have different shapes. If you think of either segment as being extended to infinity, however, you can take the other one, make a

Figure 129

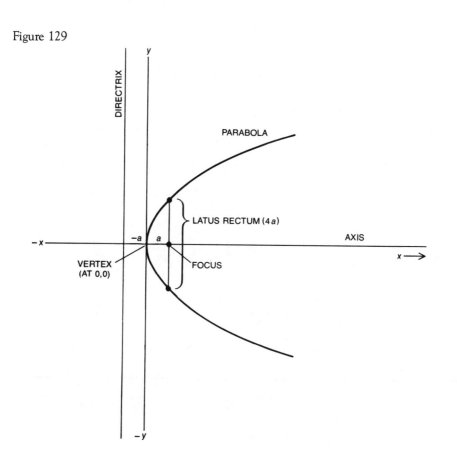

suitable change of scale and then place it somewhere on the infinite curve where it will exactly fit.

This property of varying only in size is one the parabola shares with the circle, although not with ellipses and hyperbolas. All circles are similar because single points are similar. All parabolas are similar because all pairs of a line and a point not on the line are similar. To put it another way, any directrix–focus pair will coincide with any other by a suitable dilation, translation, and rotation. Any parabola, drawn on graph paper of the right size and properly positioned, can be given any desired quadratic formula of the form $x = ay^2 + by + c$.

Figure 130

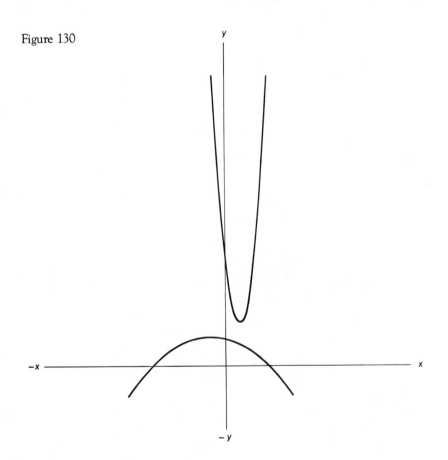

If you throw a stone horizontally, it follows a path close to a parabola, as Leonardo da Vinci conjectured in the 1490s, and Galileo proved in 1609, although he did not publish his proof until 30 years later. You can repeat one of Galileo's experiments by rolling a marble dipped in ink sideways across an inclined plane. If the plane is covered with graph paper, the path recorded by the marble will enable you to calculate the curve's parabolic formula. In actual practice the parabolic path of a projectile is slightly distorted by the earth's roundness and much more by air resistance.

In Galileo's *Dialogues Concerning Two New Sciences* he discusses at length the distorting influence of both the earth's not being flat and the air's viscosity. It is amusing to note that he discounts the deviation caused by the earth's roundness as being negligible because the range of military projectiles "will never exceed four miles."

The resistance of air to the flight of a bullet gives rise to a trajectory so resembling the curve of a breast that one of Norman Mailer's army officers, near the end of his novel *The Naked and the Dead*, sketches several pictures of the curve and muses:

> That form . . . is the fundamental curve of love, I suppose. It is the curve of all human powers (disregarding the plateau of learning, the checks upon decline) and it seems to be the curve of sexual excitement and discharge, which is after all the physical core of life. What is this curve? *It is the fundamental path of any projectile*, of a ball, a stone, an arrow (Nietzsche's arrow of longing) or of an artillery shell. It is the curve of the death missile as well as an abstraction of the life-love impulse; it demonstrates the form of existence, and life and death are merely different points of observation on the same trajectory. The life viewpoint is what we see and feel astride the shell, it is the present, seeing, feeling, sensing. The death viewpoint sees the shell as a whole, knows its inexorable end, the point toward which it has been destined by inevitable

physical laws from the moment of its primary impulse when it was catapulted into the air.

To carry this a step further, there are two forces constraining the projectile to its path. If not for them, the missile would forever rise on the same straight line. ↗ These forces are gravity and wind resistance and their effect is proportional to the square of the time; they become greater and greater, feeding upon themselves in a sense. The projectile wants to go this way ↗ and gravity goes down ↓ and wind resistance goes ←. These parasite forces grow greater and greater as time elapses, hastening the decline, shortening the range. If only gravity were working, the path would be symmetrical

it is the wind resistance that produces the tragic curve

In the larger meanings of the curve, gravity would occupy the place of mortality (what goes up must come down) and wind resistance would be the resistance of the medium . . . the mass inertia or the inertia of the masses through which the vision, the upward leap of a culture is blunted, slowed, brought to its early doom.

A jet of water from a hose also follows an almost perfect parabola. If when you water a lawn you slowly lower the angle of the hose jet from the near vertical to the near horizontal, the tops of the parabolic jets trace an ellipse, but the envelope of the jets is another parabola.

Some comets may follow parabolic paths. Comets that return periodically to the solar system move along extremely eccentric elliptical paths, but (as we have seen) the more eccentric an ellipse is, the closer it resembles a parabola. Since the parabola is a limit between the ellipse and the hyperbola, it is almost impossible to tell from observing a comet near the sun whether it is following an extremely eccentric ellipse (in which case it will return) or a parabola or hyperbola (in which case it will never return). If the path is parabolic, the comet's velocity will equal its escape velocity from the solar system. If the velocity is less, the path is elliptical; if it is more, the path is hyperbolic.

The parabola's outstanding applications in technology are based on the reflection property displayed in Figure 131. Draw a line from the focus f to any point p, and draw tangent ab to the curve at p. A line cd, drawn through p so that angle apf equals angle bpd, will be perpendicular to the directrix. It follows that if the parabola is viewed as a reflecting line, any ray of light from the focus to the curve will rebound along a path parallel to the curve's axis.

Imagine now that the parabola is rotated about its axis to generate the surface called a paraboloid. If light rays originate at the focus, the paraboloid will reflect the rays in a beam parallel to the axis. That is the principle behind the searchlight. Of course the principle also works the other way. Rays of light, parallel to the axis, shining into a concave mirror with a paraboloid surface, will all be directed toward the focus. That is the secret of reflecting telescopes, solar-energy concentrators, and microwave receiving dishes. Because large paraboloid mirrors are easier to build than transparent lenses of comparable size all giant telescopes are now of the reflecting type. Other optical devices serve to bring the image from the focus to an eyepiece or a photographic plate. As a child you may have learned how to set fire to a piece of paper by focusing the sun's rays with a glass lens. It can be done just as efficiently with a paraboloid mirror, with the paper held at the surface's focus.

If a pan of water is rotated, the surface of the water forms a paraboloid. This suggested to the physicist R. W. Wood that perhaps a reflect-

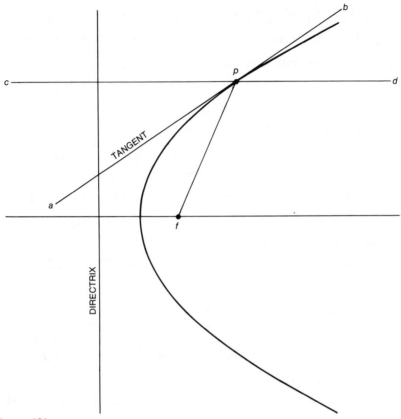

Figure 131

ing telescope could be made by rotating a large dish of mercury and exploiting the paraboloid surface as a mirror. He actually built such a telescope, but there were so many difficulties in making the surface sufficiently smooth that the idea had to be abandoned.

Assume that a paraboloid has a flat base perpendicular to its axis, so that it looks like a rounded hill. How do you calculate its volume? Archimedes found the amazingly simple formula. The volume is precisely 1.5 times that of a cone with the same circular base and the same axis.

A parabola is closely approximated by the cables that support a suspension bridge. The curve is distorted if the weight of the bridge is

not uniform or if the weight of the cables is great in relation to that of the bridge. In the latter case the curve is hard to distinguish from the one known as a catenary (from the Latin *catena*, chain). Galileo mistakenly thought the curve formed by a chain suspended at the ends was a parabola. Decades later it was shown to be a catenary, a curve that is not even algebraic because its equation contains the transcendental number *e*.

There is a curious relation between the parabola and the catenary that is not well known. If you roll a parabola along a straight line, as shown in Figure 132, top, the "locus of the focus" is a perfect catenary. Perhaps more surprising (although it is easier to prove) is what happens when two parabolas of the same size are placed with their vertexes touching and one is rolled on the other as shown in Figure 132, bottom. The focus of the rolling parabola moves along the directrix of the fixed parabola, and its vertex traces a cissoid curve!

One of the earliest problems concerning parabolas was that of "squaring" the area of a section of the curve bounded by a chord, such as the shaded region in Figure 133. The problem was first solved by Archimedes in his famous treatise *Quadrature of the Parabola*. By an

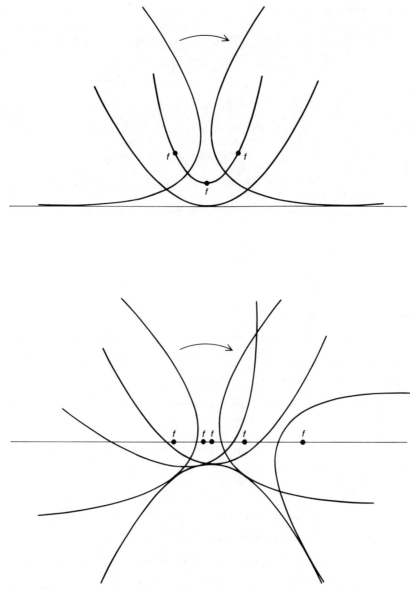

Figure 132

ingenious method of limits that anticipated the integral calculus, he
was able to prove that if you circumscribe a parallelogram as is shown
in the illustration, with its sides parallel to the parabola's axis, the area

of the parabolic segment is 2/3 that of the parallelogram. (Archimedes first guessed this by comparing the weight of the parallelogram with the weight of the segment.) Archimedes also used the parabola for an elegant way to construct the regular heptagon. Earlier geometers had exploited parabolas for the classic task of doubling the cube: constructing a cube with twice the volume of a given cube.

There are many techniques for drawing parabolas without having to plot myriads of points on a sheet of paper. Perhaps the simplest relies on a T square and a piece of string. One end of the string is attached to a corner of the T square's arm as shown in Figure 134, and the other end is attached to the parabola's focus. The string must have a length AB. A pencil point at x, pressing against the arm of the T square, keeps the string taut. As the T square slides along the directrix, the pencil moves up the side of the T square to trace the parabola's right arm. Reflecting the arrangement to the other side draws the left arm. This method was invented, or possibly reinvented, by Kepler. You can work with a right triangle or a rectangle instead of a T square and slide it

Figure 133

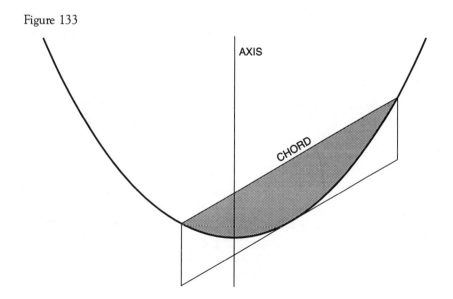

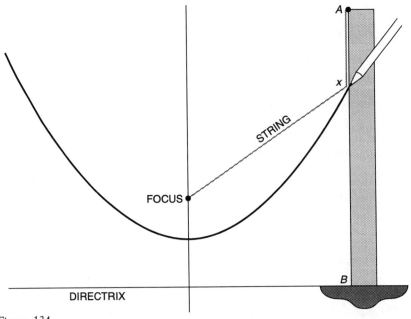

Figure 134

along the edge of a ruler. It is easy to see that the string's constant length ensures that any point on the curve is equally distant from the focus and the directrix.

Lovely parabolas can be produced more easily by paper folding. Just mark a focus point anywhere on a sheet of translucent paper, rule a directrix, and fold the sheet many times so that the line goes through the point each time. Each crease will be tangent to the same parabola, outlining the curve shown in Figure 135. If the paper is opaque, use one edge of the sheet for the directrix, folding it over to meet the point.

Familiarity with parabolas can often provide quick answers to algebraic questions. Consider, for example, this pair of simultaneous equations involving the two lucky numbers of craps:

$$x^2 + y = 7$$
$$x + y^2 = 11$$

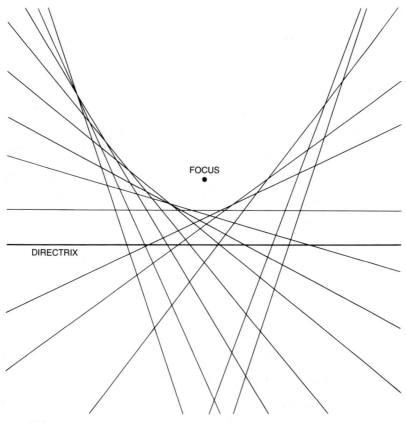

FOCUS
•

DIRECTRIX

Figure 135

It does not take long to discover that if *x* equals 2 and *y* equals 3, both equations are satisfied. Let us ask two questions:

1. Are there other integral solutions (where *x* and *y* are whole numbers, positive or negative)?

2. How many solutions are there altogether?

A more difficult problem, for which a parabola gives an answer, comes from Ronald L. Graham of Bell Laboratories, who both framed and solved it. It is published here for the first time. Imagine that you have an infinite supply of identical disks with a diameter less than 1/2 of some unit distance, say 1/10. Is it possible to put all of them in a plane, without overlapping, so that no distance between any pair of points in the disks is an integer?

Since each disk is only 1/10 in diameter, no two points on the same disk can be an integral distance apart, but it is conceivable that by a clever arrangement of disks, with their centers on a straight line, no point in one disk will be an integral distance from a point in any other disk. It is not difficult to prove that it is impossible. In fact, any arrangement of an infinity of disks on a straight line will create an infinity of pairs of disks in which an infinity of pairs of points (all exactly on the line) will be separated by integral distances.

To see how it works suppose you have placed a disk of diameter 1/10 on the line as shown by the unshaded circle in Figure 136, top. The shaded circles (marked a') are spaced with their centers a unit distance apart, extending to infinity in both directions. Clearly no second disk can be put on the line where it overlaps or touches a shaded circle; otherwise the disk will contain a point on the line that is an integral distance from a point on the line in disk a. (We assume that points on the circumference of a disk are in the disk.)

It is possible, of course, to put a second disk on the line between any pair of adjacent circles provided it does not touch or overlap either circle. For example, a second disk b can be placed as shown at the bottom of Figure 136. At once the line acquires another infinite set of circles (shaded and labeled b') at regular unit spacings, indicating that no third disk can be put where it overlaps or touches them. The same holds for additional disks. Since no more than eight disks will go without touching or overlapping in the finite space separating any pair of the first set of circles, it follows that no more than nine disks can be put on the line. A 10th disk, added anywhere, will contain an infinity of points that are integral distances from points in one of the nine previously positioned disks.

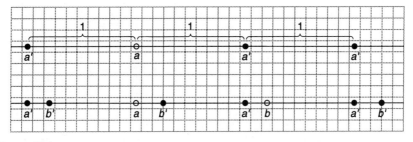

Figure 136

The proof generalizes in an obvious way to all disks smaller than 1 in diameter. If the denominator of the diameter is an integer, subtract 1 to get the maximum number of disks that can be positioned. If the denominator is not an integer, round it down to the nearest integer. Thus no disk of diameter 1 can be used. Only one disk of diameter 1/2 or diameter $1/\sqrt{2}$ can be put on the line, only two disks of diameter 1/3, only three of diameter 1/4 or diameter $1/\pi$, only four of diameter 1/4.5, and so on.

Although the problem cannot be solved by a straight line, it can be solved by a parabola.

Answers

The first problem asked for the use of a parabola to provide a quick proof of the number of solutions for the pair of equations: $x^2 + y = 7$ and $x + y^2 = 11$. The equations graph as two crossed parabolas, as shown in Figure 137. Since the parabolas intersect at just four points, there are just four solutions. Only one ($x = 2$, $y = 3$) is in integers. Even if your graphing is imprecise, you can prove that the other three solutions are not in whole numbers by testing the numbers indicated by the lattice crossings nearest the intersections. The actual numbers, all of them irrational, have the approximate values given in the illustration.

The second problem, from Ronald L. Graham, asked how an infi-

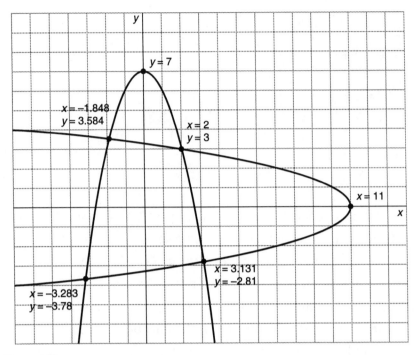

Figure 137

nite number of identical disks with diameters of less than 1, say 1/10, could be placed in the plane so that no two points in the disks are an integral distance apart. One solution is to place them with their centers on a parabola of the form $y = x^2$, with their centers at points $(1,1)$, $(3,9)$, $(9,81)$, . . ., $(3^k, 3^{2k})$, ... I lack space for Graham's unpublished proof, so I reluctantly leave it as an exercise for interested readers.

ADDENDUM

In Thomas Pynchon's well-known novel *Gravity's Rainbow*, the rainbow is the parabolic trajectory of a rocket, a symbol of the rise and fall of cultures. Pynchon strengthens the metaphor by other references to parabolas, such as the parabolic arches that led to Berlin's slums, and Germany's failed effort to construct a parabolic sound-mirror.

References

A BOOK OF CURVES. E. H. Lockwood. Cambridge University Press, 1961.

THE PARABOLA. Harold R. Jacobs in *Mathematics, a Human Endeavor: A Textbook for Those Who Think They Don't Like the Subject*. W. H. Freeman and Company, 1970.

DETERMINING THE AREA OF A PARABOLA. Jerry A. McIntosh in *Mathematics Teacher*, pages 88-91; January 1973.

SOME METHODS FOR CONSTRUCTING THE PARABOLA. Joseph E. Ciotti in *Mathematics Teacher*, Vol. 67, pages 428-430; May 1974.

GALILEO'S DISCOVERY OF THE PARABOLIC TRAJECTORY. Stillman Drake and James MacLachlan in *Scientific American*, Vol. 232, No. 3, pages 102-110; March 1975.

DO SIMILAR FIGURES ALWAYS HAVE THE SAME SHAPE? Paul G. Kumpel, Jr., in *Mathematics Teacher*, Vol. 68, No. 8, pages 626-628; December 1975.

CONSTRUCTING THE PARABOLA WITHOUT CALCULUS. Maxim Bruckheimer and Rina Hershkowitz in *Mathematics Teacher*, pages 658-662; November 1977.

19

Non-Euclidean Geometry

"Lines that are parallel
meet at Infinity!"
Euclid repeatedly,
heatedly,
    urged.
Until he died,
and so reached that vicinity:
in it he
found that the damned things
    diverged.

                    –Piet Hein, *Grooks VI*

Euclid's *Elements* is dull, long-winded, and does not make explicit the fact that two circles can intersect, that a circle has an outside and an inside, that triangles can be turned over, and other assumptions essential to his system. By modern standards Bertrand Russell could call Euclid's fourth proposition a "tissue of nonsense" and declare it a scandal that the *Elements* was still used as a textbook.

On the other hand, Euclid's geometry was the first major effort to organize the subject as an axiomatic system, and it seems hardly fair to find fault with him for not anticipating all the repairs made when David Hilbert and others formalized the system. There is no more striking evidence of Euclid's genius than his realization that his notorious fifth postulate was not a theorem but an axiom that had to be accepted without proof.

Euclid's way of stating the postulate was rather cumbersome, and it was recognized early that it could be given the following simpler form: Through a point on a plane, not on a given straight line, only one line is parallel to the given line. Because this is not quite as intuitively obvious as Euclid's other axioms mathematicians tried for 2,000 years to remove the postulate by making it a theorem that could be established on the basis of Euclid's other axioms. Hundreds of proofs were attempted. Some eminent mathematicians thought they had succeeded, but it always turned out that somewhere in their proof an assumption had been made that either was equivalent to the parallel postulate or required the postulate.

For example, it is easy to prove the parallel postulate if you assume that the sum of the angles of every triangle equals two right angles. Unfortunately you cannot prove this assumption without using the parallel postulate. An early false proof, attributed to Thales of Miletus, rests on the existence of a rectangle, that is, a quadrilateral with four right angles. You cannot prove, however, that rectangles exist without using the parallel postulate! In the 17th century John Wallis, a renowned English mathematician, believed he had proved the postulate. Alas, he failed to realize that his assumption that two triangles can be similar but not congruent cannot be proved without the parallel postulate. Long lists can be made of other assumptions, all so intuitively obvious that they hardly seem worth asserting, and all equivalent to the parallel postulate in the sense that they do not hold unless the postulate holds.

In the early 19th century trying to prove the postulate became something of a mania. In Hungary, Farkas Bolyai spent much of his life at the task, and in his youth he discussed it often with his German friend Karl Friedrich Gauss. Farkas' son János became so obsessed by the problem that his father was moved to write in a letter: "For God's sake, I beseech you, give it up. Fear it no less than sensual passions because it too may take all your time and deprive you of your health, peace of mind and happiness in life."

János did not give it up, and soon he became persuaded not only that the postulate was independent of the other axioms but also that a consistent geometry could be created by assuming that through the point an infinity of lines were parallel to the given line. "Out of nothing I have created a new universe," he proudly wrote to his father in 1823.

Farkas at once urged his son to let him publish these sensational claims in an appendix to a book he was then completing. "If you have really succeeded, it is right that no time be lost in making it public, for two reasons: first, because ideas pass easily from one to another who can anticipate its publication; and secondly, there is some truth in this, that many things have an epoch in which they are found at the same time in several places, just as the violets appear on every side in spring. Also every scientific struggle is just a serious war, in which I cannot say when peace will arrive. Thus we ought to conquer when we are able, since the advantage is always to the first comer."

János' brief masterpiece did appear in his father's book, but as it happened the publication of the book was delayed until 1832. The Russian mathematician Nikolai Ivanovitch Lobachevski had beat him to it by disclosing details of the same strange geometry (later called by Felix Klein hyperbolic geometry) in a paper of 1829. What is worse, when Farkas sent the appendix to his old friend Gauss, the Prince of Mathematicians replied that if he praised the work, he would only be praising himself, inasmuch as he had worked it all out many years earlier but had published nothing. In other letters he gave his reason. He did not want to arouse an "outcry" among the "Boeotians," by which he meant his conservative colleagues. (In ancient Athens the Boeotians were considered unusually stupid.)

Crushed by Gauss's response, János even suspected that his father might have leaked his marvelous discovery to Gauss. When he later learned of Lobachevski's earlier paper, he lost interest in the topic and published nothing more. "The nature of real truth of course cannot but be one and the same in Marcos-Vasarhely as in Kamchatka and on the moon," he wrote, resigned to having published too late to win the honor for which he had so passionately hoped.

In some ways the story of the Italian Jesuit Giralamo Saccheri is even sadder than that of Bolyai. As early as 1733, in a Latin book called *Euclid Cleared of All Blemish*, Saccheri actually constructed both types of non-Euclidean geometry (we shall come to the second type below) without knowing it! Or so it seems. At any rate Saccheri refused to believe either geometry was consistent, but he came so close to accepting them that some historians think he pretended to disbelieve them just to get his book published. "To have claimed that a non-Euclidean system was as 'true' as Euclid's," writes Eric Temple Bell (in a chapter on Saccheri in *The Magic of Numbers*), "would have been a foolhardy invitation to repression and discipline. The Copernicus of Geometry therefore resorted to subterfuge. Taking a long chance, Saccheri denounced his own work, hoping by this pious betrayal to slip his heresy past the censors."

I cannot resist adding two anecdotes about the Bolyais. János was a cavalry officer (mathematics had always been strictly a recreation) known for his swordsmanship, his skill on the violin, and his hot temper. He is said to have once challenged 13 officers to duels, provided that after each victory he would be allowed to play to the loser a piece on his violin. The elder Bolyai is reported to have been buried at his own request under an apple tree, with no monument, to commemorate history's three most famous apples: the apple of Eve, the golden apple

Paris gave Venus as a beauty-contest prize, and the falling apple that inspired Isaac Newton.

Before the 19th century had ended it became clear that the parallel postulate not only was independent of the others but also that it could be altered in two opposite ways. If it was replaced (as Gauss, Bolyai, and Lobachevski had proposed) by assuming an infinite number of "ultraparallel" lines through the point, the result would be a new geometry just as elegant and as "true" as Euclid's. All Euclid's other postulates remain valid; a "straight" line is still a geodesic, or shortest line. In this hyperbolic space all triangles have an angle sum less than 180 degrees, and the sum decreases as triangles get larger. All similar polygons are congruent. The circumference of any circle is greater than pi times the diameter. The measure of curvature of the hyperbolic plane is negative (in contrast to the zero curvature of the Euclidean plane) and everywhere the same. Like Euclidean geometry, hyperbolic geometry generalizes to three-space and all higher dimensions.

The second type of non-Euclidean geometry, which Klein names "elliptic," was later developed simultaneously by the German mathematician Georg Friedrich Bernhard Riemann and the Swiss mathematician Ludwig Schläfli. It replaces the parallel postulate with the assumption that through the point no line can be drawn parallel to the given line. In this geometry the angle sum of a triangle is always more than 180 degrees, and the circumference of a circle is always less than pi times the diameter. Every geodesic is finite and closed. The lines in every pair of geodesics cross.

To prove consistency for the two new geometries various Euclidean models of each geometry were found showing that if Euclidean geometry is consistent, so are the other two. Moreover, Euclidean geometry has been "arithmetized," proving that if arithmetic is consistent, so too is Euclid's geometry. We now know, thanks to Kurt Gödel, that the consistency of arithmetic is not provable in arithmetic, and although there are consistency proofs for arithmetic (such as the famous proof by Gerhard Gentzen in 1936), no such proof has yet been found that can be considered entirely constructive by an intuitionist [see "Constructive Mathematics," by Allan Calder; *Scientific American*, October 1979]. God exists, someone once said, because mathematics is consistent, and the Devil exists because we are not able to prove it.

The various metaproofs of arithmetic's consistency, as Paul C. Rosenbloom has put it, may not have eliminated the Devil, but they have reduced the size of hell almost to zero. In any case no mathematician today expects arithmetic (therefore also Euclidean and non-Euclidean geometries) ever to produce a contradiction. Curiously, Lewis Carroll was one of the last mathematicians to doubt non-Euclidean geometry. "It is a strange paradox," the geometer H.S.M. Coxeter has written, "that he, whose Alice in Wonderland could alter her size by eating a little cake, was unable to accept the possibility that the area of a triangle could remain finite when its sides tend to infinity."

What Coxeter had in mind can be grasped by studying M.C. Escher's *Circle Limit III*, reproduced in Figure 138. This 1959 woodcut (one of Escher's rare works with several colors in one picture) is a tessellation based on a Euclidean model of the hyperbolic plane that was constructed by Henri Poincaré. In Poincaré's ingenious model every point on the Euclidean plane corresponds to a point inside (but not on) the circle's circumference. Beyond the circle there is, as Escher put it, "absolute nothingness."

Image that Flatlanders live on this model. As they move outward from the center their size seems to us to get progressively smaller, although they are unaware of any change because all their measuring instruments similarly get smaller. At the boundary their size would become zero, but they can never reach the boundary. If they proceed toward it with uniform velocity, their speed (to us) steadily decreases, although to them it seems constant. Thus their universe, which we see

Figure 138

as being finite, is to them infinite. Hyperbolic light follows geodesics, but because its velocity is proportional to its distance from the boundary it takes paths that we see as circular arcs meeting the boundary at right angles.

In this hyperbolic world a triangle has a maximum finite area, as is shown in Figure 139, although its three "straight" sides go to infinity in hyperbolic length and its three angles are zero. You must not think of Escher's mosaic as being laid out on a sphere. It is a circle enclosing an infinity of fish—Coxeter calls it a "miraculous draught"—that get progressively smaller as they near the circumference. In the hyperbolic plane, of which the picture is only a model, the fish are all identical in size and

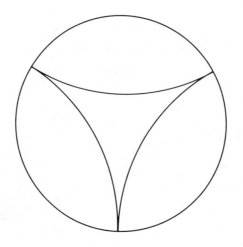

Figure 139

shape. It is important to remember that the creatures of a hyperbolic world would not change in shape as they moved about, light would not change in speed, and the universe would be infinite in all directions.

The curved white lines in Escher's woodcut do not, as many people have supposed, model hyperbolic geodesics. The lines are called equidistant curves or hypercycles. Each line has a constant perpendicular distance (measured hyperbolically) from the hyperbolic straight line that joins the arc's ends. Note that along each white curve fish of the same color swim head to tail. If you consider all the points where four fins meet, these points are the vertexes of a regular tiling of the hyperbolic plane by equilateral triangles with angles of 45 degrees. The centers of the triangles are the points where three left fins meet and three mouths touch three tails. The 45-degree angles make it possible for eight triangles to surround each vertex, where in a Euclidean tiling by equilateral triangles only six triangles can surround each vertex.

Escher and Coxeter had corresponded from the time they met in 1954, and Escher's interest in tilings of the hyperbolic plane had been aroused by the illustrations in a 1957 paper on crystal symmetry that Coxeter had written and sent to him. In a lovely article titled "The

Non-Euclidean Symmetry of Escher's Picture 'Circle Limit III'" (in the journal *Leonardo*, Vol. 12, pages 19–25; 1979) Coxeter shows that each white arc meets the boundary at an angle of almost 80 degrees. (The precise value is $2^{7/4} + 2^{5/4}$ arc secants.) Coxeter considers *Circle Limit III* the most mathematically sophisticated of all Escher's pictures. It even anticipated a discovery Coxeter did not make until five years after the woodcut was finished!

Elliptic geometry is roughly modeled by the surface of a sphere. Here Euclidean straight lines become great circles. Clearly no two can be parallel, and it is easy to see that triangles formed by arcs of great circles must have angles that add up to more than two right angles. The hyperbolic plane is similarly modeled by the saddle-shaped surface of a pseudosphere, generated by rotating a tractrix about its asymptote.

It is a misuse of the word "crank" to apply it to mathematicians who erred in thinking, before the independence of the parallel postulate was established, that they had proved the postulate. The same cannot be said of those amateurs of later decades who could not understand the proofs of the postulate's independence or who were too egotistical to try. Augustus De Morgan, in his classic compendium of eccentric mathematics, *A Budget of Paradoxes*, introduces us to Britain's most indefatigable 19th-century parallel-postulate prover, General Perronet Thompson. Thompson kept issuing revisions of his many proofs (one was based on the equiangular spiral), and although De Morgan did his best to dissuade him from his futile efforts, he was unsuccessful. Thompson also wanted to replace the tempered scale of the piano with an octave of 40 notes.

The funniest of the American parallel-postulate provers was the Very Reverend Jeremiah Joseph Callahan, president of Duquesne University in Pittsburgh. In 1931, when Father Callahan announced he had trisected the angle, *Time* gave the story sober treatment and ran his photograph. The following year Callahan published his major work,

Euclid or Einstein: A Proof of the Parallel Theory and a Critique of Metageometry (Devon-Adair, 1932), a 310-page treatise in which he ascended to heights of *argumentem ad hominem*. Einstein is "fuddled," he "has not a logical mind," he is in a "mental fog," he is a "careless thinker." "His thought staggers, and reels, and stumbles, and falls, like a blind man rushing into unknown territory." "Sometimes one feels like laughing," Callahan wrote, "and sometimes one feels a little irritated. . . . But there is no use expecting Einstein to reason."

What Callahan found so irritating was Einstein's adoption of a generalized non-Euclidean geometry, formulated by Riemann, in which the curvature of physical space varies from point to point depending on the influence of matter. One of the great revolutions brought about by relativity theory was the discovery that an enormous overall simplification of physics is obtained by assuming physical space to have this kind of non-Euclidean structure.

It is now commonplace (how astonished, and I think delighted, Kant would have been by the notion!) to recognize that all geometric systems are equally "true" in the abstract but that the structure of physical space must be determined empirically. Gauss himself thought of triangulating three mountain peaks to see if their angles added up to two right angles. It is said he actually made such a test, with inconclusive results. Although experiments can prove physical space is non-Euclidean, it is a curious fact that there is no way to prove it is Euclidean! Zero curvature is a limiting case, midway between elliptic and hyperbolic curvatures. Since all measurement is subject to error, the deviation from zero could always be too slight for detection.

Poincaré held the opinion that if optical experiments seemed to show physical space was non-Euclidean, it would be best to preserve the simpler Euclidean geometry of space and assume that light rays do not follow geodesics. Many mathematicians and physicists, including Russell, agreed with Poincaré until relativity theory changed their mind. Alfred North Whitehead was among the few whose mind was never changed. He even wrote a book on relativity, now forgotten, in which he argued for preserving a Euclidean universe (or at least one of constant curvature) and modifying the physical laws as necessary. (For a discussion of Whitehead's controversy with Einstein, see Robert M. Palter's *Whitehead's Philosophy of Science,* University of Chicago Press, 1960.)

Physicists are no longer disturbed by the notion that physical space has a generalized non-Euclidean structure. Callahan was not merely disturbed; he was also convinced that all non-Euclidean geometries are self-contradictory. Einstein, poor fellow, did not know how easy it is to prove the parallel postulate. If you are curious about how Callahan did it, and about his elementary error, see D. R. Ward's paper in *The Mathematical Gazette* (Vol. 17, pages 101–104; May 1933).

Like their cousins who trisect the angle, square the circle, and find simple proofs of Fermat's last theorem, the parallel-postulate provers are a determined breed. A recent example is William L. Fischer of Munich, who in 1959 published a 100-page *Critique of Non-Euclidean Geometry.* Ian Stewart exposed its errors in the British journal *Manifold* (No. 12, pages 14–21; Summer 1972). Stewart quotes from a letter in which Fischer accuses establishment mathematicians of suppressing his great work and orthodox journals of refusing to review it: "The university library at Cambridge refused even to put my booklet on file. . . . I had to write to the vice-chancellor to overcome this boycott."

There are, of course, no sharp criteria for distinguishing crank mathematics from good mathematics, but then neither are there sharp criteria for distinguishing day from night, life from non-life, and where the ocean ends and the shore begins. Without words for parts of continuums we could not think or talk at all. If you, dear reader, have a way to prove the parallel postulate, don't tell me about it!

Imagine a small circle around the north pole of the earth. If it keeps expanding, it reaches a maximum size at the equator, after which it starts to contract until it finally becomes a point at the south pole. In similar fashion, an expanding sphere in four-dimensional elliptical space reaches a maximum size, then contracts to a point.

In addition to the three geometries described in this chapter, there is what Bolyai called "absolute geometry" in which theorems are true in all three. It is astonishing that the first 28 theorems of Euclid's *Elements* are in this category, along with other novel theorems that Bolyai showed to be independent of the parallel postulate.

I was surprised to see in a 1984 issue of *Speculations in Science and Technology* (Vol. 7, pages 207–216), a defense of Father Callahan's proof of the parallel postulate! The authors are Richard Hazelett, vice president of the Hazelett Strip-Casting Corporation, Colchester, VT, and Dean E. Turner, who teaches at the University of North Colorado, in Greeley. Hazelett is a mechanical engineer with master's degrees from the University of Texas and Boston University. Taylor, an ordained minister in the Disciples of Christ Church, has a doctorate from the University of Texas.

It is easy to understand why both men do not accept Einstein's general theory of relativity. Indeed, they have edited a book of papers attacking Einstein. Titled *The Einstein Myth and the Ives Papers*, it was published in 1979 by Devin-Adair.

References

EUCLID'S PARALLEL POSTULATE: ITS NATURE, VALIDITY, AND PLACE IN GEOMETRI-
    CAL SYSTEMS. John William Withers. Open Court, 1905.
NON-EUCLIDEAN GEOMETRY. Robert Bonola. Open Court, 1912. Dover, 1955.
NON-EUCLIDEAN GEOMETRY. Harold E. Wolfe. Henry Holt, 1945.
THE ELEMENTS OF NON-EUCLIDEAN GEOMETRY. D. M. Y. Sommerville. Dover,
    1958.

Non-Euclidean Geometry. Stefan Kulczycki. Macmillan, 1961.

The Origin of Euclid's Axioms. S. H. Gould in *Mathematical Gazette*, Vol. 46, pages 269–290; December 1962.

Introduction to Non-Euclidean Geometry. Wesley W. Maiers in *Mathematics Teacher*, pages 457–461; November 1964.

Regular Compound Tessellations of the Hyperbolic Plane. H. S. M. Coxeter in *Proceedings of the Royal Society*, A, Vol. 278, pages 147–167; 1964.

Non-Euclidean Geometry, Fifth edition. H. S. M. Coxeter. University of Toronto Press, 1965.

The Non-Euclidean Symmetry of Escher's Picture 'Circle Limit III.' H. S. M. Coxeter in *Leonardo*, Vol. 12, pages 19–25; 1979.

Non-Euclidean Geometry. Dan Pedoe in *New Scientist*, No. 219, pages 206–207; January 26, 1981.

Euclid's Fifth Postulate. Underwood Dudley in *Mathematical Cranks*, pages 137–158. Mathematical Association of America, 1992.

Some Geometrical Aspects of a Maximal Three-Coloured Triangle-Free Graph. J. F. Rigby in *Journal of Combinatorial Theory*, Series B, Vol. 34, pages 313–322; June 1983.

The Wonderland of Poincaria. Simon Gindikin in *Quantum*, pages 21–28; November/December 1992.

Euclidean and Non-Euclidean Geometries: Development and History, Third edition. Marvin Jay Greenberg. Freeman, 1994.

20

Voting Mathematics

A variety of curious paradoxes and anomalies are involved in the process of democratic voting. The way these contradictions touch on plurality voting and the proceedings of the electoral college will be dealt with here, as will the relatively new system known as approval voting. A procedure of increasing interest to political scientists, approval voting manages to avoid many of the logical inconsistencies inherent in other voting schemes.

This chapter's discussion of the mathematics of voting is written not by me but by Lynn Arthur Steen, professor of mathematics at St. Olaf College in Northfield, MN. A former editor of *Mathematics Magazine*, Steen writes frequently on mathematical subjects and has twice been awarded the Mathematical Association of America's Lester R. Ford award for excellence in writing. He has also edited a variety of books, including *Mathematics Today: Twelve Informal Essays* (Springer-Verlag, 1978) and (with Matthew P. Gaffney) *Annotated Bibliography of Expository Writing in the Mathematical Sciences* (Mathematical Association of America, 1976).

All that follows was written by Steen, who titles his article "Election Mathematics: Do All Those Numbers Mean What They Say?"

Over past years the prospect of a three-way race for president of the United States has focused public attention on the importance of strategies for voting and on the special vagaries of the electoral college. Although the complications imposed by the electoral college are unique to presidential elections, other uncertainties imposed by three-way contests for public office are not. When the public must choose among more than two alternatives, the task of making the choice is frustratingly difficult. The source of both the difficulties and the possible solutions is to be found in the little-known mathematical theory of elections.

The social contract of a democracy depends in an obvious and fundamental way on a simple mathematical concept, namely the concept of a majority. Barring the unlikely event of a tie, in any dichotomous ballot one side or the other must receive more than half of the votes. When there are three or more choices of approximately equal strength, however, it is unlikely that such a ballot will yield a majority decision. It is primarily for this reason that many people believe the two-party system is essential to the stability of democracy in the U.S., even though that system is neither mandated nor recognized by the Constitution.

Mathematical theory and political idealism notwithstanding, quite of-

ten the public does face a choice among three or more significant alternatives. The same problem that appeared as Carter v. Reagan v. Anderson
has developed in other years. Such multiple-candidate contests are difficult
to resolve fairly if there is no clear-cut majority, but they can easily arise in
any free election. Indeed, it follows from some simple mathematics that
there are practically no positions candidates in a two-way contest can take
that are invulnerable to attack by a third or a fourth candidate.

If each issue in a two-candidate election is represented by a rating
of voter preference on a one-dimensional scale, then regardless of the
distribution of attitudes among the voters the optimal position for each
candidate is the median: the point that divides the electorate into two
camps of equal size. The same is true whether public opinion is distributed normally (so that the graph of position v. number of supporters
has a single, centered hump), is split bimodally (so that the graph has
two approximately equal humps), is skewed sharply to one side or is
divided in a highly irregular way. An example of each of these distributions, with the median marked, is given in Figure 140.

Consider a two-candidate contest in which one candidate adopts a
position a little to the left of the median and the other candidate begins
with a position at about the middle of the right half of the population.
This would be typical of a centrist candidate C running against a mod-

Figure 140

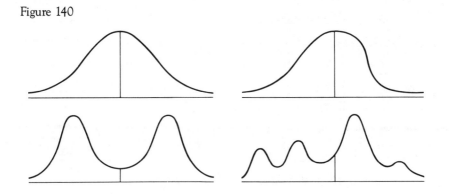

erate right-wing candidate R. In this case it is reasonable to assume that as far as this particular issue is concerned the voters whose preference lies to the left of the position held by the centrist candidate C will favor C, the voters whose preference lies to the right of candidate R will favor R, and the voters whose preference lies in between will be divided about evenly between the two candidates. Under these circumstances, in a preelection poll the centrist candidate would receive a majority of the votes.

The only way for candidate R to improve his standing in the poll (on this single issue) is to shift his position toward the middle of the distribution, to ensure that more voters will be to his right. Moving toward the center, or to the left, will always be advantageous for the right-wing candidate. Similarly, a left-leaning candidate can improve his standing with the voters by moving toward the center, or to the right. The median position is the only one that cannot be improved on by further shifting on the part of either candidate.

There is, of course, nothing very novel about this analysis. It is part of our common experience in presidential politics. Candidates representing the right or the left tend to begin distinctly to the right or to the left and then move progressively closer to the center as they attempt to appeal to a greater number of voters. The appeal of the median position in a two-candidate contest, however, is precisely what makes such a contest vulnerable to assault from either side by a third or a fourth candidate. In any contest with two candidates near the center a third candidate entering on the left or the right can always gain a plurality. Indeed, for practically any distribution of the electorate there are no positions in a two-candidate contest where at least one of the candidates cannot be beaten by a third. As is shown in Figure 141, there is always a place along the one-dimensional continuum where a new candidate can position himself to displace one or more nearby candidates.

A single issue rarely plays a deciding role in an election. Hence

Figure 141

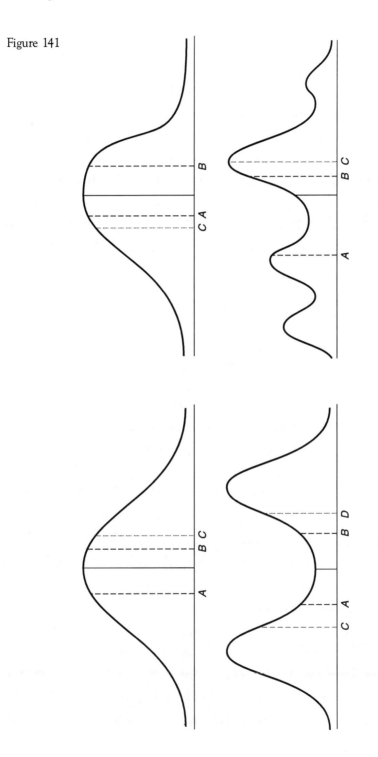

election analyses based on single issues are not very helpful, unless they can be combined to show how to design a platform that will ensure a candidate's election. Shaping a winning platform is a complex business, however, because it is possible for a platform consisting entirely of winning, or majority, planks to be defeated. The reverse side of the coin is that a majority platform can be constructed from minority planks. Hence a majority can be formed from a coalition of minorities.

To see how this paradox can arise consider the simplest possible case: a ballot to decide two unrelated, dichotomous issues, represented by resolutions A and B. In this case the voters actually have four options:

I.   Approve A and B.
II. Approve A and defeat B.
III. Defeat A and approve B.
IV. Defeat A and B.

The voters who favor both A and B would choose option I as their first choice, option IV as their fourth choice, and option II as their second or third choice, depending on whether they feel more strongly about A or about B. The voters who favor A but object to B might rank the four options in the order II, I, IV, and III (or II, IV, I, and III). In general each voter will have a preference ranking for one of the 4 × 3 × 2 × 1, or 24, possible permutations of the four available options. (The rankings are by no means equally likely; it would be hard to imagine circumstances under which many people would rank the options in the order of preference I, IV, II, III.)

Now, for the sake of simplicity suppose 500 voters (say at a party convention) are divided into three caucuses as follows: caucus X, with 150 votes, ranks the four options in the order I, II, III, IV; caucus Y, with 150 votes, ranks them II, IV, I, III, and caucus Z, with 200 votes, ranks them III, IV, I, II. In this case caucuses X and Y, with 300 votes, favor the approval of resolution A, whereas caucuses X and Z, with 350 votes, favor the approval of resolution B. Because there are different voters making up these majorities, however, the platform consisting of the planks "Approve A" and "Approve B" will be defeated by the 350-vote block of caucuses Y and Z!

This surprising phenomenon is a special case of the well-known anomaly of cyclic majorities: If three voters respectively prefer A to B to C, B to C to A, and C to A to B, then any candidate can be defeated by some other candidate by a vote of two to one in a two-candidate contest. When the issues in an election create cyclic majorities, no set of positions on the issues is invulnerable to assault by a new coalition of minorities, another factor that encourages third- and fourth-party candidates.

The accompanying diagram (Figure 142) shows how the four options from which party planks in the example must be constructed create a variety of cyclic majorities, thereby explaining how a platform

Figure 142

Options

    I. Approve resolutions A and B.
   II. Approve resolution A and defeat resolution B.
  III. Defeat resolution A and approve resolution B.
  IV. Defeat resolutions A and B.

Caucus	Policy	Votes	Order of Preference
X	Favors A strongly and favors B mildly	150	I, II, III, IV
Y	Opposes B strongly and favors A mildly	150	II, IV, I, III
Z	Opposes A strongly and favors B mildly	200	III, IV, I, II

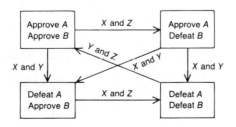

consisting of majority planks can represent the will of only a minority. The arrows joining various platforms depict voting dominance: the platform to which an arrow points will always lose to the platform at which the arrow originates in a dichotomous contest. The winning caucuses in each case appear beside the corresponding arrow. As this distribution demonstrates, any possible platform can be defeated by some other platform, and so a real convention whose divisions resemble the ones given in this example could become mired in an unending sequence of platform motions, with each motion defeating the one before.

The phenomenon of cyclic majorities is also responsible for the most famous election paradox, Kenneth J. Arrow's 1951 proof that certain generally accepted desiderata for voting schemes are logically inconsistent. If there are only two candidates, no problems arise. If three or more candidates appear on a single ballot, however, chaos reigns.

There are diverse schemes other than plurality voting for determining the winner in an election. Many were suggested by 18th-century scholars concerned about implementing the democratic ideals of the French Revolution. Although some of these proposals are so complex as to be completely impractical, several are still in common use, in particular the method of assigning points that reflect degrees of preference to the candidates in a contest (where the candidate receiving the most points is the winner) and various methods of holding runoff elections. Yet as Arrow has shown, none of these schemes—indeed, no method other than a rational benevolent dictatorship—satisfies such commonsense rules as: If A is preferred to B, and B is preferred to C, then A should be preferred to C. Cyclic majorities reduce all voting schemes to unpredictable mystery.

Another important problem with voting in three-option contests is that in many circumstances a vote for the candidate a person prefers most will increase the likelihood that the candidate he prefers least will be elected. (This dilemma was the one often seen in Anderson's candidacy. Many voters who preferred Anderson to Carter and Carter to Reagan believed most Anderson votes would be at Carter's expense.) The anomaly frequently leads thoughtful voters to what is called (depending on a voter's point of view) insincere or sophisticated voting.

If sophisticated voting is widely practiced, it can lead to a state of serious confusion where no one votes for his first choice, and so the public will is effectively camouflaged. An Anderson backer for whom Carter was a second choice might have voted for Carter instead of for Anderson in order to prevent the election of Reagan. If there were enough Anderson backers who reasoned this way, of course, some Reagan supporters might have begun to support Anderson to prevent Carter's reelection. The process of second-guessing the voting strategies of other segments of the electorate can quickly lead to an absurd hierarchy of insincerity in which the votes cast fail to reflect real preferences. Such a process, which it should be added is more a part of game theory than of classical voting theory, rarely gives a legitimate mandate to the victor.

Arrow's theorem shows there is no "perfect" voting scheme for multicandidate elections. The procedure known as approval voting, however, manages to reflect a popular will without inducing anyone to vote insincerely. In approval voting each voter marks on the ballot every candidate who meets with his approval, and the candidate who receives the most votes of approval is the winner.

With this system it is never to a voter's advantage to withhold a vote for his first choice while voting for a less preferred candidate. Indeed, if most candidates seem to have an equal chance of winning, a rational voter should vote for all the candidates he believes are above the average of those running. To vote for more candidates would give unnecessary support to individuals the voter does not endorse, whereas to vote for fewer candidates (say to vote only for one's first choice) is to withhold support from an acceptable compromise candidate and to risk victory by an unacceptable candidate.

Steven J. Brams, professor of politics at New York University, has described approval voting with the phrase "One man, n votes." It is an apt description because approval voting is merely a way of letting a person cast as many votes as he wishes, one for each acceptable candidate. It is easy to count votes that have been cast under this system, and no runoff elections are needed. For both theoretical and practical reasons approval voting is a good compromise between the single-vote ballot that encourages insincerity and the complete preference ordering whose complexity renders it useless in any practical situation.

Figure 143 shows how approval voting might compare with plurality voting, runoff voting, and point voting in an entirely hypothetical three-way contest. The number of voters supporting each of the six possible rankings of candidates are listed in the column "Total votes," and since C would receive the largest block of first-choice votes, he would win in a plurality contest. In a runoff election B would be eliminated, and A would pick up enough second-choice votes (from those

Figure 143

Order of preference	Total votes	Approval votes	
		First choice	First and second choices
A, B, C	30	20	10
A, C, B	5	5	0
B, A, C	20	10	10
B, C, A	5	5	5
C, A, B	10	5	5
C, B, A	30	20	10
Total	100	65	35

Plurality voting	Runoff voting	Point voting	Approval voting
A 35	A 35 + 20 = 55	A 35 (3) + 30 (2) + 35 = 200	A 25 + 10 + 15 = 50
B 25	C 40 + 5 = 45	B 25 (3) + 60 (2) + 15 = 210	B 15 + 10 + 20 = 45
C 40		C 40 (3) + 10 (2) + 50 = 190	C 25 + 15 + 0 = 40

who had first voted for B) to defeat C by 55 votes to 45. In the simplest system of point voting first choices are assigned three points, second choices two points, and third choices one point. Because of the large number of voters (60) for whom B is the second choice, with this voting scheme B, who was eliminated in the runoff, would be the winner.

The results of approval voting depend on whether voters find only their top choice acceptable or whether they could accept some other choices as well. (Because there are only three candidates in this example it is assumed that no one votes for all three; such a vote is legal, but it would be wasted since it would raise each candidate's total by the same amount.) In this case, with 65 voters approving only their first choice, A would receive 50 votes of approval and win the election. If some voters choose to approve two of the three candidates, however, B stands to gain most because of the large number of people who rank him as their second choice. With approval voting a shift in the number of candidates meeting the approval of even a small number of voters can easily change the outcome of the election. Hence the implementation of this voting scheme would necessitate a transformation of campaign strategies, from trying to convince voters that a candidate is the best choice to trying to convince them that he is acceptable.

In the U.S., of course, presidential elections are held by the totally different rules of the electoral college. Through most of U.S. history the electoral college has served mainly to impose a unit rule on individual states so that the winner of the popular vote in each state receives that state's entire electoral vote. According to the Constitution of the U.S., there are other significant consequences of this system that affect the outcome of three-candidate contests (in particular provisions for transferring the responsibility of deciding a presidential election from the electoral college to Congress), but here we shall examine only the consequences of the unit rule.

The most widely held view of the electoral college's unit rule has been that it favors smaller, or less populated, states, because the number of votes accorded to each state in the college is two more than its number of representatives. In relative terms these two extra votes, which represent the two senators from each state, do increase the voting strength of smaller states and diminish that of larger ones.

Paradoxically, however, the effective strength of a state in a presidential election is actually proportional to the population of the state raised to the 3/2 power. And as a result individual votes cast in the largest states are as much as three times as important as those cast in the smallest ones. This surprising conclusion is a direct consequence of elementary probability theory, and it is consistent with the spending record of candidates in recent presidential elections: candidates do devote disproportionate resources to the larger states at the expense of the smaller ones.

The "3/2 rule" is based on the assumption that candidates will generally match one another's campaign efforts in the various states. (Comparison of candidates' allocations of time and money in recent election campaigns shows that the assumption is entirely realistic.) The reasoning begins with the obvious: Each candidate seeks to maximize his expected electoral vote, which is the sum over all 50 states of the product of each state's electoral vote and the probability that the candidate will win a majority in that state. By expressing this relation in the form of an equation and taking into account candidates' tendencies to match one another's campaign efforts from state to state it can be shown that the optimal way to maximize the expected electoral vote is to allocate campaign resources approximately in proportion to the 3/2 power of the electoral vote of each state. Thus although California has about four times the electoral vote of Wisconsin (45 compared with 11), the 3/2 rule would suggest that candidates should devote $4^{3/2}$, or 8, times more resources to California than to Wisconsin.

Another way to understand why larger states gain power rather than lose it in electoral-college politics is to examine the likelihood that any particular vote may be decisive in swinging the state for or against a particular candidate. This measure of decisiveness is the traditional way of gauging the power of an individual voter. What is needed is a measure of the average number of votes necessary to reverse the result of an election in each state.

Calculations show that the decision power of an individual in a state with v electoral votes (to be cast as a unit in the electoral college) is proportional to $\sqrt{v}$. Since the power of a state in the electoral college is magnified by the number of electoral votes cast by the state, the contribution of each state to the presidential decision is approximately proportional to v times $\sqrt{v}$, or $v^{3/2}$.

In order to gauge the relative voting power of individuals in different states the large-state bias created by the 3/2 rule must be weighed against the small-state bias of the two-senator electoral-college bonus. The significance of an individual's vote, instead of being equal for all voters, is determined by the individual's share of his state's power, and as shown in Figure 144 the different states' powers are decidedly un-

Figure 144

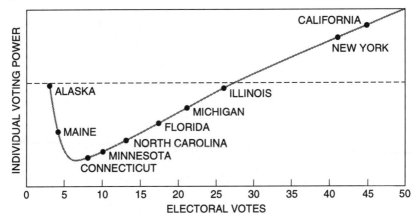

equal. (The broken line on the graph marks the hypothetical even distribution of power.)

Elections will always remain a matter of passion more than of logic,
based on belief more than on reason. As these examples demonstrate,
however, the mathematics of elections can have subtle and unexpected
consequences. As in many other realms of human experience, naive
expectations can be shattered by simple mathematical structures dis-
guised as paradoxes and anomalies.

References

THEORY OF VOTING. Robin Farquharson. Yale University Press, 1969.

THE PRESIDENTIAL ELECTION GAME. Steven J. Brams. Yale University Press,
1978.

APPROVAL VOTING: A 'BEST BUY' METHOD FOR MULTI-CANDIDATE ELECTIONS.
Samuel Merrill in *Mathematics Magazine*, Vol. 52, pages 98–102; March
1979.

APPROVAL VOTING: A PRACTICAL REFORM FOR MULTICANDIDATE ELECTIONS. Steven
J. Brams in *National Civic Review*, Vol. 68, pages 549–553, 560; November 1979.

ONE CANDIDATE, ONE VOTE: WILL APPROVAL VOTING REVOLUTIONIZE 20TH-
CENTURY ELECTIONS? Steven J. Brams, in *Archway*, pages 11–14; Winter
1981.

21

A Toroidal
Paradox
and Other
Problems

1   A Poker Puzzle

As every poker player knows, a straight flush (Figure 145, left) beats four of a kind (Figure 145, right).

How many different straight flushes are there? In each suit a straight flush can start with an ace, a deuce, or any other card up to a 10 (the ace may rank either high or low), making 10 possibilities in all. Since there are four suits, there are four times 10, or 40, different hands that are straight flushes.

How many different four-of-a-kind hands are there? There are only

Figure 145

13. If there are 13 four-of-a-kind hands and 40 straight flushes, why does a straight flush beat four of a kind?

2   The Indian Chess Mystery

Raymond M. Smullyan's long-awaited collection of chess problems was published in 1979 by Knopf with the title *The Chess Mysteries of Sherlock Holmes*. Just as there has never been a book of logic puzzles quite like Smullyan's *What Is the Name of This Book?*, so there has never been a book of chess problems as brilliant, original, funny, and profound as this one.

Although a knowledge of chess rules is necessary, as Smullyan says in his introduction, the problems in the book actually lie on the borderline between chess and logic. Most chess problems deal with the future, such as how can White move and mate in three. Smullyan's problems belong to a field known as retrograde analysis (retro analysis for short), in which it is necessary to reconstruct the past. This can be done only by careful deductive reasoning, by applying what Smullyan calls "chess logic."

Sherlock Holmes would have had a passion for such problems, and his enthusiasm would surely have aroused the interest of Dr. Watson, particularly after Watson had learned from Holmes some of the rudiments of chess logic. Each problem in Smullyan's book is at the center of a Sherlockian pastiche narrated by Watson in his familiar style. Some of the problems are so singular that it is difficult to believe they have answers. In one, for example, Holmes proves that White has a mate in two but that it is not possible to show the actual mate. In another problem Holmes shows that in the days when chess rules allowed a promoted pawn to be replaced by a piece of the opposite color, a position could arise in which it is impossible to decide if castling is legal even when all the preceding moves are known.

In the second half of the book Holmes and Watson set sail for an island in the East Indies where they hope to find a buried treasure by combining cryptography with retrograde chess analysis. Their first adventure takes place on the ship. Two men from India have been playing a game with pieces that are colored red and green instead of the usual black and white or black and red.

The players have temporarily abandoned the game to stroll around the deck when Holmes and Watson arrive on the scene. The position of the game is shown in Figure 146. Several chess enthusiasts are studying the position and trying to decide which color corresponds to White, that is, which side had made the first move.

"Gentlemen," says Holmes, "it turns out to be quite unnecessary to guess about the matter. It is *deducible* which color corresponds to White."

In retro-chess problems it is not required that one side or the other play good chess, only that they make legal moves. Your task is to decide which color moved first and prove it by ironclad logic.

3   Redistribution in Oilaria

"Redistributive justice" is a phrase much heard these days in arguments among political philosophers. Should the ideal modern industrial state tax the rich for the purpose of redistributing wealth to the poor? Yes, says Harvard philosopher John Rawls in his influential book *A Theory of Justice*. No, says his colleague (they have adjoining offices) Robert Nozick in his controversial defense of extreme libertarianism, *Anarchism, State and Utopia*. It is hard to imagine how two respected

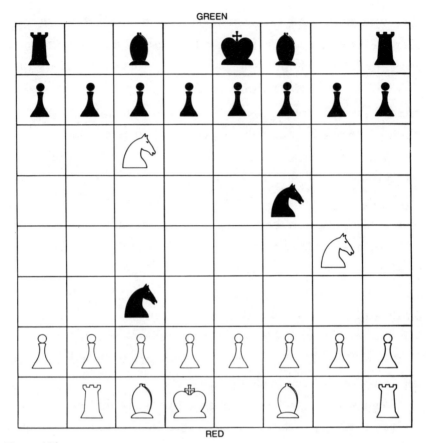

Figure 146

political theorists, both believing in democracy and free enterprise, could hold such opposing views on the desirable powers of government.

The Sheik of Oilaria, in this problem devised by Walter Penney of Greenbelt, MD, has never heard of Rawls or Nozick, but he has proposed the following share-the-wealth program for his sheikdom. The population is divided into five economic classes. Class 1 is the poorest, class 2 is the next-poorest, and so on to class 5, which is the richest. The plan is to average the wealth by pairs, starting with classes 1 and 2, then 2 and 3, then 3 and 4, and finally 4 and 5. Averaging means that the total wealth of the two classes is redistributed evenly to everyone in the two classes.

The Sheik's Grand Vizier approves the plan but suggests that averaging begin with the two richest classes, then proceed down the scale instead of up.

Which plan would the poorest class prefer? Which would the richest class prefer?

4 Fifty Miles an Hour

A train goes 500 miles along a straight track, without stopping, completing the trip with an average speed of exactly 50 miles per hour. It travels, however, at different speeds along the way. It seems plausible that nowhere along the 500 miles of track is there a segment of 50 miles that the train traverses in precisely one hour.

Prove that this is not the case.

5 A Counter-Jump "Aha!"

Draw a five-by-six array of spots on a sheet of paper, then rule a line as is shown in Figure 147 to divide the array into two triangular halves of 15 spots each. On the spots above the line [*shown black*] place 15 pennies or any other kind of small object.

Figure 147

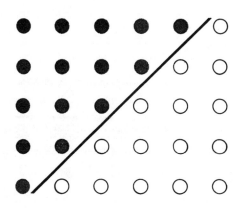

The task is to move all the pennies from above the line to the spots below the line. Each move is a jump of one counter over an adjacent counter to an unoccupied spot immediately beyond it on the other side. Jumps may be to the left or the right and up or down but not diagonal. For example, as a first move the penny at the fourth spot on the top row may jump to the top white spot, or it may jump down to the third spot from the top of its column. All the jumps are like the jumps in checkers except that they are confined to horizontal and vertical directions and the jumped pieces are not removed.

We are not concerned with transferring the pennies to the white spots in a minimum number of moves, only with whether the transfer can be made at all. There are three questions:

A. Can the task be accomplished?

B. If a penny is removed from a black spot, can the 14 pennies that remain be jumped to white spots?

C. If two pennies are removed from black spots, can the remaining 13 be jumped to white spots?

This new problem was devised recently by Mark Wegman of the Thomas J. Watson Research Center of the International Business Machines Corporation. It is of special interest because all three questions can be answered quickly by an "Aha!" insight well within the grasp of a 10-year-old.

6   A Toroidal Paradox

Two topologists were discussing at lunch the two linked surfaces shown at the left in Figure 148, which one of them had drawn on a paper napkin. You must not think of these objects as solids, like ropes or solid rubber rings. They are the surfaces of toruses, one surface of genus 1 (one hole), the other of genus 2 (two holes).

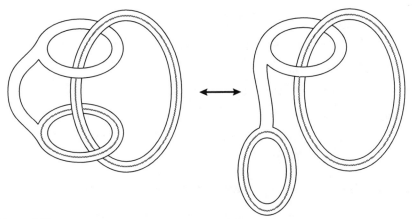

Figure 148

Thinking in the mode of "rubber-sheet geometry," assume that the surfaces in the illustration can be stretched or shrunk in any desired way provided there is no tearing or sticking together of separate parts. Can the two-hole torus be deformed so that one hole becomes unlinked as is shown at the right in the illustration?

Topologist X offers the following impossibility proof. Paint a ring on each torus as is shown by the black lines. At the left the rings are linked. At the right they are unlinked.

"You will agree," says X, "that it is impossible by continuous deformation to unlink two linked rings embedded in three-dimensional space. It therefore follows that the transformation is impossible."

"But it doesn't follow at all," says Y.

Who is right? I am indebted to Herbert Taylor for discovering and sending this mystifying problem.

Answers

1   A Poker Puzzle

The poker puzzle is answered when we consider the fact that a hand with four identical values always has a fifth card. For each four of a kind there are 48 different fifth cards. Consequently there are 48×13, or 624, different poker hands containing four of a kind, compared with 40 hands that are straight flushes. It is therefore much less likely that you will be dealt a straight flush, and for this reason a straight flush beats four of a kind. The problem was contributed by M. H. Greenblatt to *Journal of Recreational Mathematics* (Vol. 5, No. 1, page 39; January 1972).

2   The Indian Chess Mystery

Here is how Raymond Smullyan proves that Green made the first move:

> Red is now in check, hence Green moved last. It remains to determine who moved first, which can be done by figuring whether an odd or an even number of moves have been made.
>
> The rook on b1 has made an odd number of moves; the other three rooks have each made an even number of moves (possibly zero). The Red knights have collectively made an odd number of moves, since they are on squares of the same color, and the Green knights have collectively made an even number of moves. [A knight changes square color on each move.] One king has made an even number of moves (possibly zero), and the other king an odd number. The bishops and pawns have never moved, and both queens were captured before they ever moved. So the grand totality is odd. Thus Green moved first. Hence Green is White and Red is Black.

3   Redistribution in Oilaria

Surprising as it may first seem, both the richest and the poorest classes in Oilaria would prefer pair averaging from the top down. Those in the richest class would prefer to be averaged with the next-richest class before the latter is reduced in wealth by averaging. Those in the poorest class would prefer being averaged with the next-poorest class after the latter has been increased in wealth by averaging.

An example will make this clear. Assume that the wealth of the five classes is in the proportions 1 : 3 : 4 : 7 : 13. Averaging from the bottom up changes the proportions to 2 : 3 : 5 : 9 : 9. Averaging from the top down changes the proportions to 3 : 3 : 5 : 7 : 10.

Robert Summers, professor of economics at the University of Pennsylvania, sent the following comments on the Oilaria problem:

> Of course, it is trivial to show that both the low and high income people would prefer the Grand Vizier's variant to the original income redistribution plan proposed by the Sheik. It should be enough simply to observe that both the poor and the rich would like their incomes averaged with others having as large an income as possible. The Grand Vizier's variant—averaging from above—adjusts upward the income of the next-to-the-bottom class before it is combined with the bottom class; similarly, the income of the next-to-the-top class is combined with the top class before it is lowered through the averaging process. Whether or not another income class, the second or next to last or any one in between, would like the redistribution from either or both averaging methods depends upon the particular income distribution. The fact that the so-called Lorenz curves cross reflects this indeterminacy.
>
> There are several interesting questions that go beyond the exercise as you gave it: (1) If the two redistribution schemes were repeated over and over again, how would the limiting distributions obtained by averaging from above and below compare? Is one better than the other from the standpoint of any particular class? Repeating the arithmetic over and over again with any starting income distribution will suggest that if either averaging process is repeated indefinitely, a completely equal—egalitarian—distribution will eventually result. Any one with an initial income below the average will be pleased with the iterated redistribution done either way; anyone above the average will dislike either redistribution. This can be shown in general by drawing upon some theorems about doubly stochastic matrices. (2) Suppose a random pair

of persons is selected from Oilaria's population, and each person in the pair is given half of the total income of the two. If this process were repeated to cover every possible pairing, what would the resulting distribution be like? Suppose the random pairings, independent from trial to trial, occurred an infinite number of times. What would the limiting distribution be? Strong conjecture: Again an egalitarian distribution would result.

4   Fifty Miles an Hour

Divide the 500-mile track into 10 segments of 50 miles each. If any segment is traversed in one hour, the problem is solved, and so it must be assumed that traversing each segment takes either less than an hour or more than an hour. It then follows that somewhere along the track there will be at least one pair of adjacent segments, one (call it A) traversed in less than an hour and the other (call it B) traversed in more than an hour.

Imagine an enormous measuring rod 50 miles long that is placed over segment A. In your mind slide the rod slowly in the direction of segment B until it coincides with B. As you slide the rod the average time taken by the train to go the 50 miles covered by the rod varies continuously from less than an hour (for A) to more than an hour (for B). Therefore there must be at least one position where the rod covers a 50-mile length of track that was traversed by the train in exactly one hour.

For a more technical analysis of the problem, in terms of a jogger who averages a mile in eight minutes and runs an integral number of miles, see "Comments on 'Kinematics Problem for Joggers,' by R. P. Boas, in the *American Journal of Physics* (Vol. 42, page 695, August 1974).

5   A Counter-Jump "Aha!"

The Aha! insight that solves the counter-jumping puzzle is to color nine spots black as is shown in Figure 149. It is obvious that, no

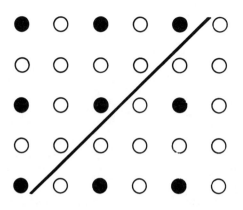

Figure 149

matter how many jumps are made, a penny on any black spot can go only to another black spot.

There are six black spots above the line and only three below it. Therefore, by the pigeonhole principle, there must be three pennies above the line that have nowhere to go below the line. The task of moving all the pennies to spots below the line cannot be accomplished unless at least three pennies, on three black spots, are removed from the top triangular array. Remove any three such pennies and the transfer of the remaining 12 is a simple task.

Benjamin L. Schwartz wrote to point out that the proof of impossibility holds even if the allowed moves are extended to include diagonal jumps, or bishop moves along diagonals without jumping. David J. Abineri sent a different proof of impossibility also based on the pigeonhole principle.

Number the columns 1, 2, 3, 4, 5, 6. Spots in columns 1, 3, and 5 are black. No matter how a penny jumps, orthogonally or diagonally, it must remain on a black spot. At the start nine pennies occupy black spots. But there are only six black spots above the line.

6   A Toroidal Paradox

Figure 150 shows how a continuous deformation of the two-hole

Figure 150

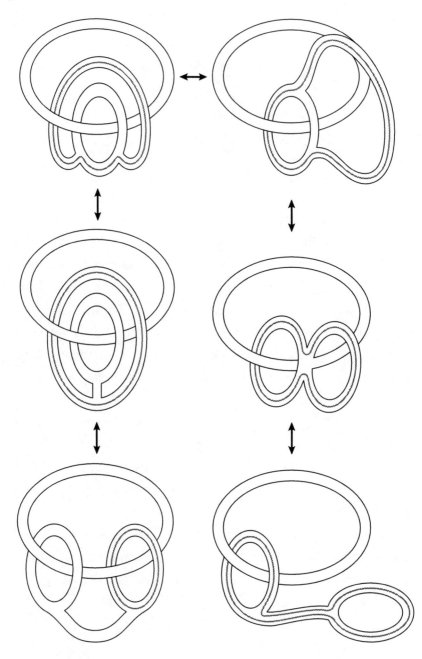

torus will unlink one of its holes from the single-hole torus. The argument for the impossibility of this task fails because if a ring is painted around one hole (as is shown by the black line), the ring becomes distorted in such a way that after the hole is unlinked the painted ring remains linked through the one-hole torus.

For a mind-boggling selection of similar problems involving linked toruses, see Herbert Taylor's article "Bicycle Tubes Inside Out," in *The Mathematical Gardner* (1981), edited by David Klarner.

22

Minimal
Steiner
Trees

No tree in all the grove but has its
   charms,
Though each its hue peculiar.

> —William Cowper, *The Task*,
> *Book 1: The Sofa*

In graph theory, the study of structures formed by joining points with lines, a tree is a connected network of line segments that includes no circuits. A circuit is a closed path that allows one to travel along a connected network from a given point back to itself without retracing any lines. It follows that any two points on a tree are joined by a unique path. Trees are extremely important in graph theory, and they have endless applications in other branches of mathematics, particularly probability theory, operations research, and artificial intelligence.

Suppose a finite set of n points are randomly scattered about in the plane. How can they be joined by a network of straight lines that has the shortest possible total length? The solution to the problem has practical applications in the construction of such networks as roads, power lines, pipelines, and electrical circuits. If no new points are allowed to be added to the original set, the shortest network connecting them is called a minimal spanning tree. It is easy to see the network must be a tree: if it included a circuit, one could shorten it at once by removing a line from the circuit.

There are many ways to construct a minimal spanning tree. The simplest is known as a greedy algorithm, because at each step it bites off the most desirable piece. It was published in 1956 by Joseph B. Kruskal, now at the AT&T Bell Laboratories. First find two points that are closer together than any other two and join them. If more than one pair of points are equally close, choose any such pair. Repeat the procedure with the remaining points in such a way that joining a pair never completes a circuit. The final result is a spanning tree of minimal length.

A minimal spanning tree is not necessarily the shortest network spanning the original set of points. In most cases a shorter network can be found if one is allowed to add more points. For example, suppose you want to join the three points defining the corners of an equilateral triangle. Two sides of the triangle make up a minimal spanning tree. The spanning tree can be shortened by more than 13 percent by adding an extra point at the center and then making connections only between the center point and each corner (see Figure 151, top). Each angle at the center is 120 degrees.

A less obvious example is the minimal network spanning the four corners of a square. You might suppose one extra point in the center would give the minimal network, but it does not. The shortest network requires two extra points (see Figure 151, right). Again all the angles

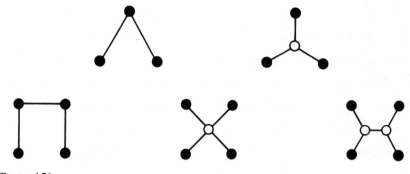

Figure 151

around the extra points in the network are 120 degrees. The network with one extra point in the center has length $2\sqrt{2}$, or about 2.828. The network with two extra points reduces the total length to $1+\sqrt{3}$, or about 2.732.

One of the first mathematicians to investigate such networks was Jacob Steiner, an eminent Swiss geometer who died in 1863. The extraneous points that minimize the length of the network locally are now called Steiner points. (I shall describe what is meant by "locally" below.) It has been proved that all Steiner points are junctions of three lines forming three 120-degree angles. A tree with Steiner points is called a Steiner tree. Although adding Steiner points can reduce the length of the spanning tree, a Steiner tree is not always the shortest network spanning the original set of points. When it is, it is called a minimal Steiner tree.

Minimal Steiner trees are almost always shorter than minimal spanning trees, but the reduction in length may depend on the length of the original spanning tree. It has been conjectured that for any given set of points in the plane, the length of the minimal Steiner tree cannot be less than $\sqrt{3}/2$, or about .866, times the length of the minimal spanning tree; the result has been proved, however, only for three, four, and five points. Just as a set of points can have more than one minimal spanning tree, so it can have more than one minimal Steiner tree, although of course all minimal Steiner trees for a given set of points have the same length. A Steiner tree can have at most $n - 2$ Steiner points, where n is the number of points in the original set.

Many simple Steiner trees can be found empirically by a simple analog device you can build. Two parallel sheets of Plexiglas are joined by perpendicular rods that correspond to the points to be spanned in a given network. Drill holes in the sheets, insert the rods and immerse the entire assembly in a soap solution of the kind used for making bubbles. When the assembly is lifted out of the solution, a soap film forms surfaces that span the rods. Because such surfaces shrink to minimal area, the pattern formed by the film when it is viewed from above is a Steiner tree.

Such a device can find the minimal Steiner tree for the corners of a rectangle (see Figure 152). The tree can take either of two forms, one of them a 90-degree rotation of the other. By blowing on the film you can make it jump from one pattern to its rotated form. Similarly, the device can model the minimal Steiner tree for the five points at the corners of a regular pentagon. For the six corners of a regular hexagon (and all higher regular polygons) extra Steiner points are of no help. The minimal spanning network is simply the perimeter of the polygon with one edge removed.

Even in these simple cases, however, one must be wary of the soap-film computer. For example, if the four points in the given network mark the corners of a rectangle a trifle wider than it is high, the film

Figure 152

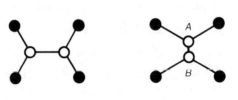

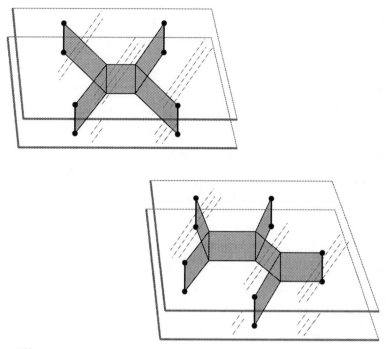

Figure 153

can stabilize in one of two patterns (see Figure 153). Both are Steiner trees, but only the one at the left is minimal. As the rectangle widens, the vertical line AB in the nonminimal pattern on the right becomes shorter. The line shrinks to a point when the vertical side of the rectangle is 1 and the base is $\sqrt{3}$, and for all wider rectangles only the minimal Steiner tree is stable. The tree at the right is said to be locally minimal. In other words, if you think of the lines as being elastic bands anchored at their ends to the four corner pegs, any slight shifting of the extra points will increase the length of the tree.

Given the simplicity of Kruskal's greedy algorithm for the construction of minimal spanning trees, one might suppose there would be correspondingly simple algorithms for finding minimal Steiner trees. Such, alas, is not the case. The task is part of a special class of "hard" problems known in computer science as NP-complete. When the num-

ber of points in a network is small, there are known algorithms for
finding Steiner trees in a reasonably short time. As the number of
points grows, however, the computing time needed increases at a rapidly accelerating pace. Even for a relatively small number of points it
can be thousands or even millions of years. Most mathematicians believe no efficient algorithm exists for constructing minimal Steiner
trees that connect arbitrary points in the plane.

Imagine, however, that the points are arranged in a regular lattice
of unit squares, like the points at the corners of the cells of a checkerboard. Is there a "good" algorithm for finding a minimal Steiner tree
spanning the points of such regular patterns?

The question occurred to me several years ago when I thought of
the following problem. What is the length of the minimal Steiner tree
that joins the 81 points at the corners of a standard checkerboard?
Henry Ernest Dudeney, England's greatest puzzle maker, and his American counterpart Sam Loyd were both fond of puzzles based on checkerboard patterns. I checked all their books carefully, but they had not
considered the problem. Indeed, I could find no evidence it had ever
been posed before, let alone solved.

When I tried to solve the problem, I was surprised by its complexity. Although I could not prove it, it seemed obvious that the minimal
Steiner tree would be constructed by joining many replicas of the regular four-point tree. The four-point tree has no name; let us call it X
because in working on Steiner-tree problems for rectangular lattices, an
X is easier to draw than the full tree. The difficulty in solving such
problems is that it is hard to know where to place the Xs. It is easy to
place them so as to make a Steiner tree, but it is not so easy to make
the tree minimal.

I finally convinced myself that the checkerboard puzzle has a unique
answer, although I could not prove it (see Figure 154, center). I call
it the conjectured solution for the order-9 array, where the order is

Figure 154

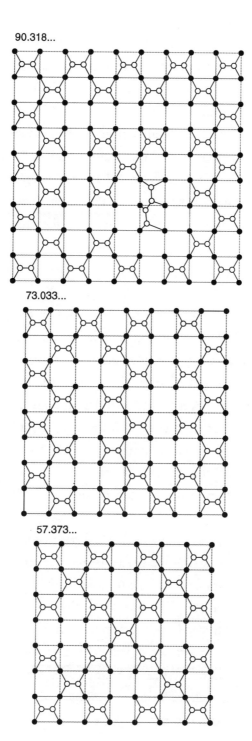

the number of points on the side of the square. Because the length of the line segments that make up each X is $1+\sqrt{3}$, it is easy to determine the total length of the tree: $26\sqrt{3}+28$, or about 73.033. Although it seemed I had found a new puzzle, I suspected that in the growing mathematical literature on Steiner trees there must surely be a paper describing a simple algorithm for finding minimal Steiner trees on rectangular lattices. I was encouraged by knowing that many problems involving paths through points in the plane, which are hard when the points are arbitrary, become trivial when the points form regular lattices.

The traveling salesman problem is a notorious example. What is the shortest path allowing a salesman to visit each of n towns once and only once and return to the starting town? When the points are arbitrary, the task is NP-complete, and no efficient algorithm for solving it is known. But when the points are placed at the corners of squares and packed into a rectangular lattice, the problem is absurdly easy. If a rectangular array of m-by-n points includes an even number of points, the minimal path has length $m \times n$. If the array includes an odd number of points, the path has length $m \times n + \sqrt{2} - 1$ (see Figures 155 and 156). I fully expected that the task of spanning points in such arrays by minimal Steiner trees would be equally trivial. I could not have been more wrong.

My first step was to send the checkerboard problem to my friend Ronald L. Graham, a distinguished mathematician at Bell Laboratories. I also asked him to direct me to a paper that might answer such questions. To my amazement, it turned out that the only relevant paper was one Graham himself had coauthored in 1978 with Fan R. K. Chung, also of Bell Laboratories. Titled "Steiner Trees for Ladders," it showed how to construct minimal Steiner trees for 2-by-n rectangular arrays of points, as well as for other kinds of 2-by-n "ladders." Aside from these special cases, nothing seemed to be known about how to

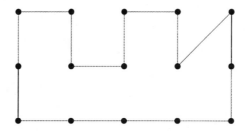

Figure 155

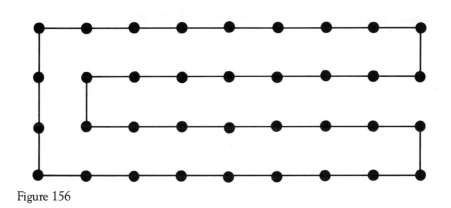

Figure 156

find minimal Steiner trees for rectangular arrays when the number of points on each side is greater than 2.

The more Graham and Chung considered the matter, the more it intrigued them. On and off for better than a year they have been seeking an algorithm for the general case, but without success. Chung has recently been lecturing on the topic, and she and Graham plan eventually to write a paper on their progress.

Their best results are shown along with my checkerboard solution in Figures 154 and 157. Some of the trees have more than one minimal solution. Incredibly, only the pattern for the order-2 square lattice has been proved to be minimal. (There is a proof in Problem 73 of the book *100 Problems in Elementary Mathematics*, by Hugo Steinhaus.) Even the seemingly trivial order-3 pattern has eluded proof, although it

Figure 157

43.980...

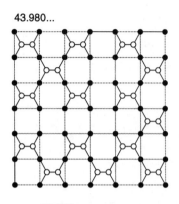

31.984...

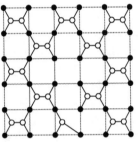

22.124...

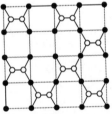

13.660...

7.464...

would yield to brute-force methods carried out by computer. Graham and Chung firmly believe all their trees are minimal, but in the absence of proofs there may still be room for improvements.

It would be interesting to know whether soap film will solve the square lattices of order 3 and order 4. If it does, how far up the scale will soap film continue to find minimal trees? What happens when Plexiglas sheets, joined by 81 rods in the checkerboard pattern, are dipped into the soap solution and then lifted out? Will the film generate Steiner trees spanning all 81 rods? If it does, what is the probability the tree will be minimal? Perhaps some venturesome readers will carry out these experiments.

The order-6 square lattice is the smallest one from which an unexpected solution springs. When I worked on this forest of trees (sets of disconnected trees are known as forests to graph theorists), my order-6 pattern had length $11\sqrt{3}+13$, or about 32.053. I almost fell out of my chair when I saw the shorter tree found by Graham and Chung. The little three-point tree in their pattern has length $(1+\sqrt{3})/\sqrt{2}$, and so the total length of their network is $[(1+\sqrt{3})/\sqrt{2}]+[11\times(1+\sqrt{3})]$, or about 31.984. It beautifully illustrates the kind of surprises—the "hue peculiar" of Cowper's epigraph—that lie in wait for anyone who tries to climb the ladder of square arrays in search of minimal solutions.

If you look closely, you will note that only squares of orders that are powers of 2 (2, 4, 8, and so on) have trees made entirely of Xs. Graham and Chung have proved an even more general result: a rectangular array can be spanned by a Steiner tree made up entirely of Xs if and only if the array is a square and the order of the square is a power of 2. Their clever proof, based on mathematical induction, is still unpublished. The unique spanning pattern generalizes in an obvious way to all squares whose order is a higher power of 2.

Space does not allow me to provide examples of the best-known

patterns for nonsquare rectangular arrays, for which Graham and Chung
have many curious results and conjectures. I close by giving the best
Steiner tree they have found for the order-22 square (see Figure 158). It
includes a pattern bounded by six points on two squares, which does
not match the familiar X. The six-point pattern also appears in the
Steiner tree of order 10, and its length is

$$\sqrt{11 + 6\sqrt{3}},$$

or about 4.625. The total length of the tree is approximately 440.021.

Figure 158

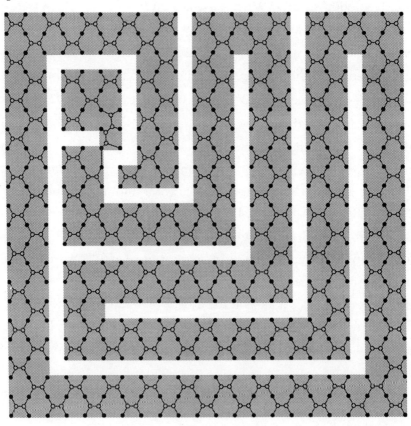

Answers

Readers were challenged to find a Steiner tree shorter than 32.095 units that would span a 4-by-9 array. No one yet has improved on the result obtained by Ronald L. Graham and Fan R. K. Chung of AT&T Bell Laboratories (see Figure 159). Its length is

$$7[1+\sqrt{3}]+$$
$$[(([([3(2+\sqrt{3})]-2)^2+1]^{1/2})/2]+$$
$$[(([([5(2+\sqrt{3})]-2)^2+1]^{1/2})/2],$$

or approximately 32.094656. . . .

ADDENDUM

Two major breakthroughs relating to MSTs (minimal Steiner trees) have occurred since this chapter was written in 1986. In 1968 two Bell Labs mathematicians, H. O. Pollok and E. N. Gilbert, conjectured that the ratio of the length of an MST to the length of the minimal spanning tree for the same set of points is at least $\sqrt{3}/2 = .866\ldots$, a savings in length of about 13.4 percent. This is the ratio for the two kinds of trees that join the corners

Figure 159

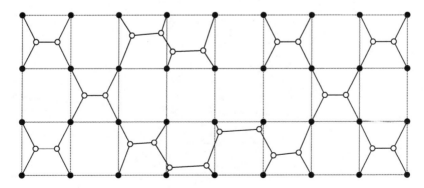

of an equilateral triangle. (See the 1968 paper by Gilbert and Pollok.) In 1985 Ronald Graham and his wife Fan Chung raised the lower bound of the ratio to .8241. The proof was so horrible, Graham said, that he urged those interested *not* to look up their paper.

The problem was important enough to Bell Labs, where finding shorter networks is an obvious cost saving, for Graham to offer $500 to anyone who could prove the $\sqrt{3}/2$ conjecture. The prize was won in 1990 by two Chinese mathematicians, Ding Zhu Du, then a postgraduate student at Princeton University, and Frank Hwang, of Bell Labs. (See their 1992 paper.)

A simplex is a regular polyhedron, in any dimension, with a minimum number of sides, such as the 3-space tetrahedron. MSTs are known only for simplexes through five dimensions. (See the 1976 paper by Chung and Gilbert.) Calculating them for higher dimensions is far from solved. The MST for the corners of a unit cube is shown in Figure 160. Its length is 6.196. . . .

Hwang and Du, in their 1991 paper, study MSTs on isometric (equilateral triangle) lattice points.

The other breakthrough, by five Australian mathematicians, was a complete solution to finding MSTs for both square and rectangular lattices of points in a matrix of unit squares. (See the 1995 Research Report by M. Brazil and his four associates.) In a 1996 paper by Brazil and five associates they confirmed the unpublished proof by Graham and Chung about the form of MSTs for sets of points at the vertices of a $2^k \times 2^k$ square lattice.

Figure 160

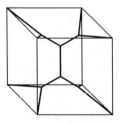

There is a growing literature on minimal *rectilinear* Steiner trees—trees with only horizontal and vertical lines. They have important applications to electrical circuit design. See Dana Richard's "Fast Heuristic Algorithms for Rectilinear Steiner Trees," in *Algorithmica*, Vol. 4, 191–207; 1989.

In a lecture on Steiner trees by Graham, which I had the pleasure of attending, he included the following points:

Jacob Steiner made no contributions to the theory of Steiner trees except to get his name attached to them. The points were earlier called Fermat points, but their existence was known even before Fermat's time.

The first proof that finding Steiner trees for *n* points is NP-complete was in the 1977 paper by Graham and two Bell Labs colleagues Michael Garey and David Johnson. Also NP-complete is the problem of calculating the exact length of a minimal spanning tree. Intuitively it seems as if the greedy algorithm would make this easy. It is not easy because the points may not be at integer coordinates on the plane. As such points increase in number, calculating the exact length of the tree rapidly becomes more difficult.

References

STEINER MINIMAL TREES. E. N. Gilbert and H. O. Pollok in *SIAM Journal of Applied Mathematics*, Vol. 16, No. 1, pages 1–29; January 1968.

STEINER TREES FOR THE REGULAR SIMPLEXES. Fan Chung and E. N. Gilbert in *Bulletin of the Institute of Mathematics Academy Sinica*, Vol. 4, pages 313–325; 1976.

ON STEINER MINIMAL TREES WITH RECTILINEAR DISTANCE. F. H. Hwang in *SIAM Journal of Applied Mathematics*, Vol. 30, pages 104–114; 1976.

THE COMPLEXITY OF COMPUTING STEINER MINIMAL TREES. M. R. Garey, R. L. Graham, and D. S. Johnson in *SIAM Journal of Applied Mathematics*, Vol. 32, pages 835–859; 1977.

STEINER TREES FOR LADDERS. Fan Rong K. Chung and R. L. Graham in *Annals of Discrete Mathematics*, Vol. 2, pages 173–200; 1978.

SMART SOAP BUBBLES CAN DO CALCULUS. Dale T. Hoffman in *The Mathematics Teacher*, Vol. 72, No. 5, pages 377–385, 389; May 1979.

A NEW BOUND FOR EUCLIDEAN STEINER MINIMAL TREES. Fan Chung and Ronald Graham in *Annals of the New York Academy of Sciences*, Vol. 440, pages 328–346; 1985.

THE SHORTEST-NETWORK PROBLEM. Marshall Bern and Ronald Graham in *Scientific American*, Vol. 260, pages 84–89; 1989.

STEINER TREES ON A CHECKERBOARD. Fan Chung, Martin Gardner, and Ron Graham, in *Mathematics Magazine*, Vol. 62, pages 83–96; April 1989.

STEINER MINIMAL TREES ON CHINESE CHECKERBOARDS. F. K. Hwang and D. Z. Du in *Mathematics Magazine*, Vol. 64, pages 332–339; December 1991.

THE STEINER TREE PROBLEM. F. K. Hwang, D. S. Richards, and P. Winter in *Annals of Discrete Mathematics*, Vol. 53, Amsterdam, 1992.

OPTIMAL STEINER POINTS. Regina B. Cohen, in *Mathematics Magazine*, Vol. 65, pages 323–329; December 1992.

A PROOF OF THE GILBERT–POLLOK CONJECTURE ON THE STEINER RATIO. D. Z. Du and F. K. Hwang in *Algorithmica*, Vol. 7, pages 121–135; 1992.

MINIMAL STEINER TREES FOR THREE-DIMENSIONAL NETWORKS. R. Bridges in *The Mathematical Gazette*, Vol. 78, pages 157–162; July 1994.

FULL MINIMAL STEINER TREES ON LATTICE SETS. M. Brazil, J. H. Rubinstein, J. F. Weng, N. C. Wormald, and D. A. Thomas. *Research Report 14*, Department of Electrical Engineering, University of Melbourne, Australia, pages 1–40; 1995.

MINIMAL STEINER TREES FOR RECTANGULAR ARRAYS OF LATTICE POINTS. M. Brazil, J. H. Rubinstein, D. A. Thomas, J. F. Weng, and N. C. Wormald, *Research Report 24*, University of Melbourne, Australia, pages 1–28; 1995.

MINIMAL STEINER TREES FOR $2^k \times 2^k$ SQUARE LATTICES. M. Brazil, T. Cole, J. H. Rubinstein, D. A. Thomas, J. F. Weng, and N. C. Wormald, in *Journal of Combinatorial Theory*, Series A, Vol. 73, pages 91–109; January 1996.

23

Trivalent
Graphs,
Snarks,
and
Boojums

W hen the following chapter ran as a column in *Scientific American* (April 1976) the famous four-color map theorem was still an open question. A number of distinguished mathematicians were on record as believing the theorem false.

The simplest flawed proof rests on a common confusion of the four-color theorem with a much simpler theorem that is easy to prove. No more than four regions can be drawn on the plane so that every pair share a common border segment. It is tempting to think this can lead to a proof of the four-color theorem. The great British puzzle expert Henry Dudeney, in *Modern Puzzles* (1926), actually published just such a "proof." Where did it go wrong? Because it is conceivable that a map with a large number of regions, say many thousands, might not be four-colorable even though nowhere on it would be a spot where five regions mutually shared borders. Any attempt to four-color such a map would lead to a spot where two regions of the same color came together. Of course one could always eliminate the spot by backtracking and altering colors, only to discover that as coloring continued the troublesome spot would turn up somewhere else. To prove the four-color theorem it would be necessary to specify a precise coloring procedure guaranteed to four-color the entire map.

Because the four-color map theorem was finally proved in 1976, I left this column out of a previous book collection which otherwise should have included it. We now know that what I called a Boojum— a Snark that does not contain the Petersen graph—cannot exist. However, interest in Snarks has persisted, with numerous papers about them still appearing in the journals. For this reason I decided to reprint the column here, deleting paragraphs that are no longer of any interest. Snark theory has become much too complicated to cover adequately in my brief addendum. Interested readers must consult the articles listed in the references.

Dozens of conjectures, seemingly unrelated to maps, are equivalent to the four-color map theorem in the sense that if you settle any one of them, you settle the four-color problem. Is it always possible to slice corners from a convex polyhedron until every face is a polygon with a number of sides that is a multiple of 3? If you can do that, the four-color conjecture is true! For instance, by truncating four corners of a cube in such a way that no two corners are diagonally opposite, you can produce a solid with four triangular faces and six hexagonal ones. If you can find a convex polyhedron that cannot be properly transformed by truncation, you will have found a solid whose skeleton will produce a map disproving the four-color theorem.

A much easier way to search for a countermap is to look for a graph (a set of points called vertexes joined by lines called edges) that has the following properties:

1. *It is connected.* (It is all in one piece.)

2. *It is planar.* (It can be drawn on the plane with no edge intersections.)

3. *It has no bridge (or isthmus).* A bridge is an edge such that if it is removed, the graph falls apart into two disconnected pieces.

4. *It is trivalent.* (Three edges meet at every vertex.)

5. *It is not three-colorable.* (The edges cannot be colored with three colors, one to an edge, so that all three colors meet at every vertex.)

To explain this more fully, let us go back to a paper published in 1880 by Peter G. Tait, a mathematical physicist at the University of Edinburgh. Tait and others showed how easily any map can be transformed into a trivalent map with the same coloring properties. If a vertex has more than three edges, draw a small circle around it. Erase what is inside, and also erase one of the circle's arcs (see Figure 161). The vertex of *n* edges is replaced by an extension of one region, now

Figure 161

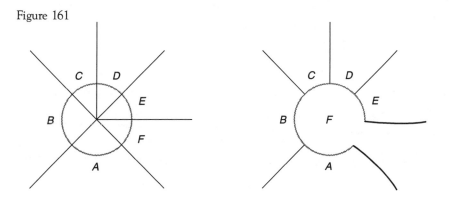

surrounded by $n - 2$ trivalent vertexes. It is obvious that any coloring of the regions will color the original map. We do not have to worry about a vertex with only two edges because it is just a spot on a border and can be removed. In brief, any map can be changed to a network of trivalent forking lines, producing a trivalent map, and if the trivalent map can be four-colored, so can the original map. In addition Tait was able to prove that if the regions of a planar trivalent map can be four-colored, the edges of its graph can be three-colored, and vice versa.

The equivalence of the two colorings is evident from the following procedures. Assume that the regions of a trivalent map are colored with A, B, C, and D. Label each edge with a letter that is the "sum" of the regions on each side, using the following addition table:

$$A + B = B$$
$$A + C = C$$
$$A + D = D$$
$$B + C = D$$
$$B + D = C$$
$$C + D = B$$

The result is a three-coloring of the edges. To go from edge-coloring to region-coloring assume that the edges of a trivalent graph are colored with B, C, and D. Label any region A. From A take any path that goes from face to face. When you cross an edge, label the new region with the "sum" of the edge and the last region visited. Use the same addition table as before if the two letters differ, and call the new region A if they are the same. The result is a four-coloring of the regions.

Tait believed all trivalent graphs are three-colorable (and therefore all maps are four-colorable), with the exception of two kinds of trivalent graphs. One consists of trivalent maps with bridges. Three simple graphs of this type are shown in Figure 162. The two loops in the first graph make it obviously uncolorable, and it is almost as trivially obvious that

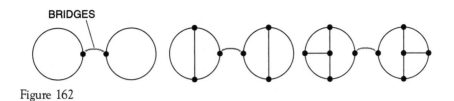

BRIDGES

Figure 162

the other two graphs also cannot be three-colored. Such graphs cannot, of course, be those of any legitimate map because the bridge would divide the outside region—a connected region if the map is on a sphere—from itself. (When any map is four-colored on the plane, the "outside" must always be treated as a region.) The bridge would be an absurd border: if you crossed it, you would still be in the same region.

The other class of uncolorable trivalent maps are all nonplanar (impossible to draw on the plane without at least one intersecting edge). The simplest example, known as the Petersen graph, is shown in Figure 163. The form on the left is the one usually seen in textbooks. Rufus Isaacs of Johns Hopkins University, an applied mathematician noted for his work on game theory, prefers the one on the right. It is drawn with fewer strokes, and all the vertexes except one are on the outside, where

Figure 163

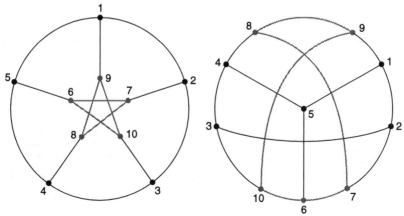

they can be used for "hooking" the graph to other graphs in a manner to be explained below. By tracing each graph along its edges, taking the spots in order, you can easily verify that the two are topologically identical. The inner star is colored to make the isomorphism more obvious.

What can we conclude from all of this? If an uncolorable trivalent graph exists that is equivalent to a map disproving the four-color theorem, it must be trivalent, planar and bridge-free. No such graph has been found, and the reader would be ill-advised to search for one. On the other hand, as Isaacs recently discovered, the search for nonplanar graphs that cannot be "Tait-colored" (three-colored) can be a delightful pastime. Isaacs' book *Differential Games* (Robert E. Krieger, 1975) has provided material for several of my columns. The results of his search for uncolorable trivalent graphs are reported in his fascinating paper "Infinite Families of Nontrivial Trivalent Graphs Which Are Not Tait Colorable," in *The American Mathematical Monthly*, Vol. 82, No. 3, pages 221–239; March 1975.

By nontrivial Isaacs means primarily a graph without a bridge. It is so easy to attach a bridge to any graph and render it uncolorable that we can ignore all graphs with bridges and concentrate on trivalent graphs without them. Isaacs also considers trivial any graph that has a "digon" (multiple edges joining two points), "triangle" (or quadrilateral) because these are trivial features that can be added to or removed from any uncolorable graph without altering its uncolorable property.

To avoid the constant use of "nontrivial uncolorable trivalent" it would be helpful to have a single term for it. My first thought was the acronym NUT, suggesting that such graphs, like buried nuts, are hard to find. Indeed, as Isaacs puts it, anyone who searches for them "will be vividly impressed with the maddening difficulty of finding" a single one. NUT, however, implies that the search is a bit nutty when in fact it is serious mathematical business. The problem of defining and classifying all nontrivial uncolorable trivalent graphs is as worthy of being tackled as proving the four-color-map theorem. If it is ever solved, and one can prove that all such graphs are planar, the four-color-map problem will also be solved.

I propose calling nontrivial uncolorable trivalent graphs Snarks. A trivalent graph is a network of forking paths, and the person who tries to prove that it is uncolorable is certainly pursuing "with forks and hope" like the mad Snark-hunting crew in Lewis Carroll's immortal nonsense ballad. We know that Snarks are difficult to find, and that there is an exceedingly rare and dangerous variety called a Boojum. In our terminology the Boojum is none other than the planar Snark: the trivalent graph that explodes the four-color conjecture by providing a countermap. If anyone discovers a Boojum, he and the graph are instantly translated into hyperspace. This may explain why the four-color problem remains open. Was Judge Crater an amateur mathematician?

The Petersen graph, first published in 1898, not only is the smallest possible Snark but also is (as W. T. Tutte, a great expert on graph theory, has shown) the only Snark with as few as 10 spots (points) on its hide. It is hard to believe, but more than half a century passed before the second Snark (with 18 spots) was discovered by Danilo Blanuša, who published it in 1946. Two years later Blanche Descartes (a pseudonym used by Tutte) published a 210-spot Snark. Not until 1973 was the fourth Snark (50 spots) published by G. Szekeres.

The main outcome of Isaacs's Snark expedition was the discovery of two infinite sets of Snarks. One set includes the graphs of Blanuša, Descartes, and Szekeres. Isaacs calls them BDS graphs to honor the three mathematicians on whose work he based his own. The graphs are formed by hooking together graphs previously known to be uncolorable, and also by hooking on other arbitrary graphs. Blanuša was not aware of it, but his graph can be obtained by joining two Petersen graphs in the manner shown in Figure 164. Remove any two nonadjacent edges from the Petersen graph on the left, then remove the

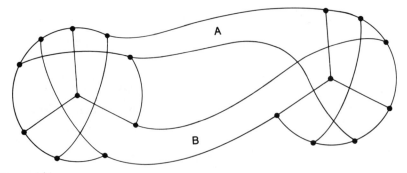

Figure 164

three adjacent edges and their two vertexes shown missing on the Petersen graph on the right. Hook the two graphs together as indicated. The pairs of edges at A can be crossed or not, and the same is true of the pair at B. Moreover, at A or B (or both) one can hook in arbitrary graphs without destroying colorability. Any two uncolorable graphs can be joined in this way, with or without additional arbitrary graphs, to produce an infinity of Snarks. Szekeres' graph is formed by hooking together five Petersen graphs. Descartes's graph is a combination of Petersen graphs with inserted nonagons. Readers interested in these constructions can consult Isaacs' paper for details.

The second infinite set of Snarks found by Isaacs is shown in Figure 165. The first graph is the Petersen graph with a trivial "triangle" of three spots substituted for the central vertex. By increasing the number of large petals in the series 3, 5, 7, 9, . . . one obtains an infinite set of flower Snarks with spots in the series 12, 20, 28, 36, ... Isaacs provides a simple visual proof that all flower Snarks are uncolorable.

All trivalent graphs, by the way, necessarily have an even number of spots. If the number is $2n$, the number of edges is $3n$, and if the graph is three-colorable, there are n edges of each color.

Isaacs also discovered one Snark (30 spots) of a wild variety that does not belong to either the BDS or the flower sets. He calls it the double-star (see Figure 166). Of course, the double-star Snark, as well

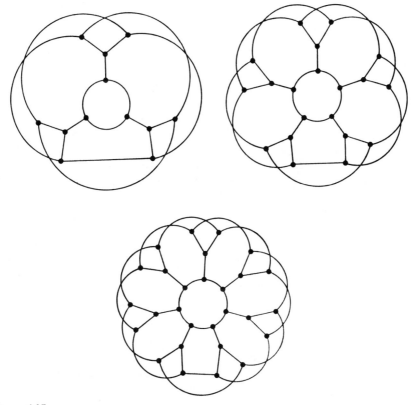

Figure 165

as any flower graph, can be hooked to BDS graphs, or the flower graphs can be hooked to one another. The combinatorial possibilities are endless, and complex BDS graphs can be drawn in such a way that it is extremely difficult to sort out their component parts.

To introduce readers to the excitement of Snark hunting, four simple trivalent graphs that are three-colorable are presented in Figure 167. Readers are invited to see how quickly they can three-color any one graph or all four graphs. The bottom graph is drawn in canonical form; that is, so all its points are on a straight line. After some practice at three-coloring the reader may want to try the more difficult task of proving that the Petersen graph or any of the other Snarks is uncolorable. To do this

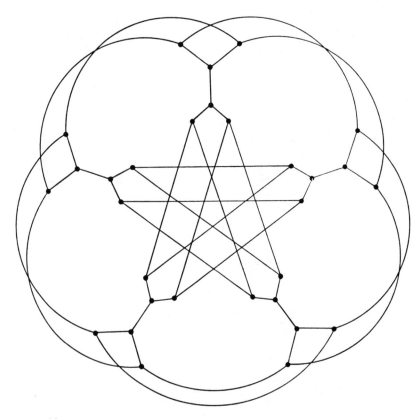

Figure 166

you must test all possible ways of three-coloring, and that can be time-consuming, particularly if you are not using an efficient procedure.

The following backtrack algorithm, which is not given in Isaacs' paper, is the one he has found most useful, both for three-coloring trivalent graphs and for proving their uncolorability. I shall describe it in pencil-and-paper terms (see Figure 168), although it can be made even more efficient by working with extremely large graphs and small numbered counters.

1. Draw a large picture of the graph in ink. Let 1, 2, and 3 stand for the colors. All the labeling should be done with a soft lead pencil because there may be many erasures.

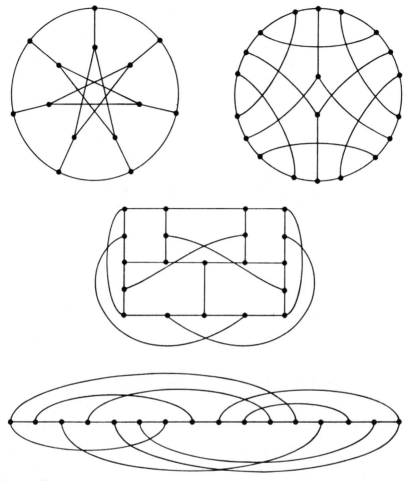

Figure 167

2. Pick any vertex and label its three forking paths, 1, 2, and 3. It does not matter how you distribute the numbers. They merely stand for different colors, so that there is no loss of generality by permuting them.

3. Move to any adjacent vertex. Its two unlabeled edges can be labeled in two ways. In our example the upper edge must be 1 or 3. Label it 1, putting a bar over the numeral to indicate that it is a free choice. Add subscript 1 to show that it is your first free choice. We call this the step number.

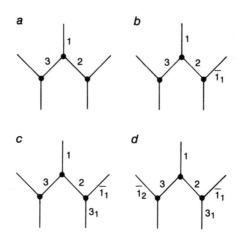

Figure 168

4. Label all edges that are determined by the first free choice. In our example there is only one. It gets a 3. Add the same step number (subscript 1). No bar goes over the 3 because the labeling is forced.

5. Move to another adjacent vertex where there is a free choice. As before, the first decision gets a bar over it, but now the subscript is 2. This shows that it is your second free choice.

6. Continue in this way until the entire graph is labeled or you encounter a contradiction—a forced choice that puts two edges of the same color at a vertex. When that occurs, at step n, erase all labels having subscript n. It is wise to leave the erasure of the barred label to the last.

7. Make the alternative choice at step n. This time, however, no bar is put over the label. Why? Because it is no longer a free choice. It was forced by the contradiction that resulted from the previous choice at step n. It gets the subscript $n - 1$. This shows that you were forced to backtrack one step. In other words, the new step has now become part of the previous step, so that it gets the previous step's number.

8. Keep repeating the procedure. If the graph is colorable, you will eventually color it. If it is not, you will keep encountering contradictions

that force backtracking. Subscripts will become fewer only to increase again. If there is a contradiction after all edges have acquired labels with no bars, then the graph is uncolorable. You have found a Snark.

"The task is simpler," Isaacs writes in a letter, "the more omni-scient we are in foreseeing the consequences of each step." After a while numerous dodges will occur to the experienced Snark hunter. Figure 169 shows some useful coloring tips. For example, a digon (a two-sided cycle) on a path can be skipped because clearly the same color is forced on both sides. Similarly, a triangle can be treated like a single vertex because the three edges leading to it (as you can easily

Figure 169

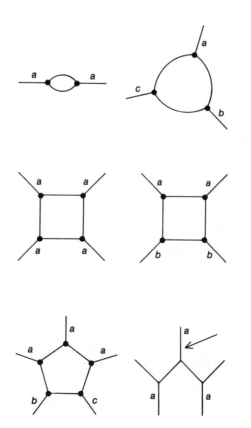

prove) must have three colors. Squares can be simplified by remembering that either the four edges leading to it are all the same color or two adjacent edges are one color and the other two are another. Pentagons are simplified by remembering that, of the five leading edges, three adjacent edges must be the same, the other two colors being on the remaining pair.

It is good, Isaacs advises, to hunt for forks of the kind depicted in the last graph in Figure 168. If two edges are the same color, as shown, the edge marked with the arrow must also be the same color. Finding forced labels of these kinds can eliminate unnecessary backtracking.

Every known Snark, Isaacs tells us, contains at least one Petersen graph. This means that by erasing certain edges and removing spots from edges that remain, you are left with a structure topologically identical with the Petersen graph. It does not mean that the Petersen graph is a subgraph. Subgraphs have to correspond, point for point, edge for edge, to a portion of a graph. Although subgraphs are contained within graphs, not all contained graphs are subgraphs. Indeed, the Petersen graph cannot be a subgraph of a trivalent graph, because if you add an edge to any vertex of the Petersen graph, it raises the order of the vertex to 4.

Figure 170 shows one way to remove five edges (shown as broken lines) and 10 spots from the flower Snark to leave a Petersen graph. The graph is numbered and colored to correspond with the graph in Figure 163. The presence of this graph inside the Snark proves it is not planar, because no planar graph can contain a nonplanar one. There is a famous theorem that states all nonplanar graphs (not necessarily trivalent ones) must contain either the complete graph for five points or the six-point "utilities" graph. Along similar lines Tutte has conjectured that all Snarks contain Petersen graphs. If that is true, the four-color theorem is also true and Boojums do not exist.

In our Carrollian terminology Petersen graphs are the "bathing machines" that Snarks are so fond of, and that Carroll says every Snark

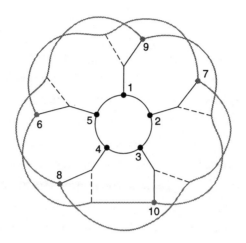

Figure 170

> . . . constantly carries about,
> And believes that they add to the beauty
>   of scenes—
> A sentiment open to doubt.

Finding bathing machines inside Snarks is not always easy. You might try looking for one in the other two flower Snarks and inside the double-star. Perhaps you will become sufficiently intrigued to embark on a Snark-hunting expedition of your own. Try drawing and testing a variety of trivalent graphs. Your ability to color them will improve rapidly with practice, and you will become impressed with how difficult it is to find genuine Snarks. If you bag any with more than 10 spots (a Petersen graph) and fewer than 18 (a Blanuša graph), I should like to see their picture. No Snarks are known with spots numbering between 10 and 18. The first flower Snark (12 spots) does not count because the triangle at its center makes it a trivial variation of the Petersen graph.

Hold on a minute! I have just finished sketching a fantastic trivalent graph with 50 projections that stick out like feathers. There are no intersections. It just might be a Bo. . . .

ADDENDUM

When I asked readers for examples of Snarks with 12, 14, or 16 points, I failed to make clear that they must be nontrivial Snarks. A Snark is considered trivial if it contains loops (cycles) that are digons (double edges), triangles, or quadrilaterals, or if an arbitrary subgraph is added by means of "bridges." Multiple edges can of course be changed to a single edge. A triangle (three-sided loop) can be contracted to a single point to make a smaller Snark, and a quadrilateral (four-sided loop) can be replaced by two edges.

One trivial way to add a subgraph is to sever an edge, as shown at the left of Figure 171, and use bridges to add a subgraph. Another way uses bridges to replace any point with an arbitrary subgraph as shown on the right of the same figure. Snarks remain Snarks under both types of alteration. Isaacs' original paper bans such trivial changes. Many readers sent Snarks with 12, 14, and 16 points but which were trivial. Since my column appeared it has been proved that nontrivial Snarks with 12, 14, and 16 points cannot exist. Snarks have been constructed with 18, 20, 22, 26, 28, 30 points, and higher.

Roman Nedela and Martin Skoviera, in their 1996 paper, show that a 24-point nontrivial Snark (they call such Snarks "irreducible") does not exist. They raise the open question: For which even numbers n greater than 10 does there exist an irreducible Snark of order n?

There are many ways to draw a Petersen graph. Jan Mycielski prefers the graph at the left of Figure 172. On the right is shown how Isaacs drew a

Figure 171

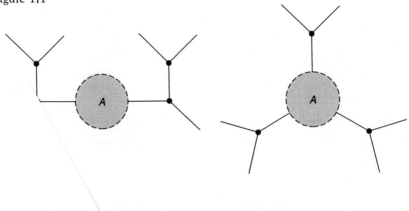

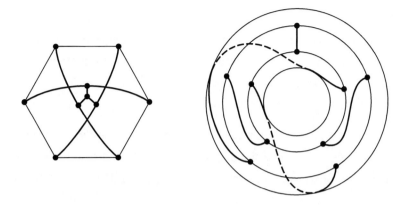

Figure 172

Petersen graph on a torus. It requires five colors. A proof that the Petersen graph requires four colors is so simple and elegant that I present it here in the form given in the paper by Amanda Chetwynd and Robin Wilson.

Color the outside edges of the Petersen graph with three colors, *a, b, c,* as shown in Figure 173. This forces the colors of the five spokes. The two dotted edges now contradict the assumption that the Petersen graph is four-colorable. One dotted line must be colored *a,* forcing the other dotted line to have a fourth color.

Before Isaacs wrote his 1975 paper only four nontrivial Snarks were known. Several infinite families of Snarks have since been discovered

Figure 173

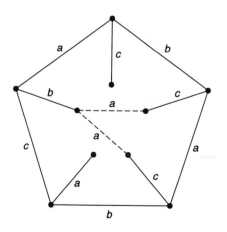

(Figure 174 shows an example of a nontrivial Snark with 22 points). The major task still open is to classify all types of Snarks in a systematic way. Is there, for instance, a way to define "prime Snarks" so that every Snark can be constructed from prime Snarks?

The following two conjectures also remain unsolved:

1. Does every Snark contain a Petersen graph? As we have seen, if this can be proved it would provide a simple proof of the four-color map theorem.

2. Does every Snark contain a cycle of 5 or 6 points? This is known as the "girth conjecture." A graph's girth is the smallest cycle (if any) that it contains. A Snark cannot have a cycle of 2, 3, or 4 points, so it must have a cycle of at least 5. Put another way, no Snark has a girth of 7 or greater.

Figure 174

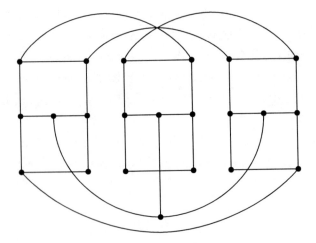

R̃ṛ̃ences

IES OF NONTRIVIAL TRIVALENT GRAPHS WHICH ARE NOT TAIT
Rufus Isaacs in *American Mathematical Monthly*, 82, pages
·h 1975.

BIFAMILY OF NON-TAIT-COLORABLE GRAPHS. Rufus Isaacs,

Technical Report 263. Department of Mathematical Sciences, Johns Hopkins University, November 1976.

THE CONSTRUCTION OF SNARKS. U. A. Celmins and E. R. Swart. Research Report CORR 79-18. University of Waterloo, Canada, 1979.

SNARKS AND SUPERSNARKS. Amanda G. Chetwynd and Robin J. Wilson in *The Theory and Application of Graphs*. Wiley, pages 215-224, 1981.

ON THE CONSTRUCTION OF SNARKS. John J. Watson in *Ars Combinatoria*, 16-B, pages 111-123, 1983.

DECOMPOSITION OF SNARKS. Peter J. Cameron, Amanda G. Chetwynd, and John J. Watkins in *Journal of Graph Theory*, 11, pages 13-14, Spring 1987.

SNARKS. John J. Watkins in *Annals of the New York Academy of Sciences*, 576, pages 606-622, 1989.

A SURVEY OF SNARKS. John J. Watkins and Robin J. Wilson in *Graph Theory, Combinatorics, and Applications*, Vol. 2. Wiley, pages 1129-1144, 1991.

DECOMPOSITION AND REDUCTIONS OF SNARKS. Roman Nedela and Martin Skoviera in *Journal of Graph Theory*, 22, pages 253-279, 1996.

A CYCLICALLY CONNECTED 6-EDGE CONNECTED SNARKS OF ORDER 118. Martin Kochol in *Discrete Mathematics*, 161, pages 297-300, 1996.

Index